# FIELDS

VINCENT J. HYDE

**BALBOA.**PRESS
A DIVISION OF HAY HOUSE

Copyright © 2021 Vincent J. Hyde.

All rights reserved. No part of this book may be used or reproduced by any means, graphic, electronic, or mechanical, including photocopying, recording, taping or by any information storage retrieval system without the written permission of the author except in the case of brief quotations embodied in critical articles and reviews.

Balboa Press books may be ordered through booksellers or by contacting:

Balboa Press
A Division of Hay House
1663 Liberty Drive
Bloomington, IN 47403
www.balboapress.com.au
AU TFN: 1 800 844 925 (Toll Free inside Australia)
AU Local: 0283 107 086 (+61 2 8310 7086 from outside Australia)

Because of the dynamic nature of the Internet, any web addresses or links contained in this book may have changed since publication and may no longer be valid. The views expressed in this work are solely those of the author and do not necessarily reflect the views of the publisher, and the publisher hereby disclaims any responsibility for them.

The author of this book does not dispense medical advice or prescribe the use of any technique as a form of treatment for physical, emotional, or medical problems without the advice of a physician, either directly or indirectly. The intent of the author is only to offer information of a general nature to help you in your quest for emotional and spiritual well-being. In the event you use any of the information in this book for yourself, which is your constitutional right, the author and the publisher assume no responsibility for your actions.

Any people depicted in stock imagery provided by Getty Images are models, and such images are being used for illustrative purposes only.
Certain stock imagery © Getty Images.

Print information available on the last page.

ISBN: 978-1-9822-9024-5 (sc)
ISBN: 978-1-9822-9025-2 (e)

Balboa Press rev. date: 05/27/2021

# CONTENTS

Author's Notes ................................................................... vii
Explanation of the Use of Upper Case ............................... ix
Quote Relevant To The Book ............................................. xi

| | | |
|---|---|---|
| Chapter 1 | Introduction ............................................... | 1 |
| Chapter 2 | The Black Hole Field .................................. | 6 |
| Chapter 3 | The Galaxy Field ........................................ | 11 |
| Chapter 4 | The Star Field ............................................. | 14 |
| Chapter 5 | The Planetary and Moon Fields ................. | 21 |
| Chapter 6 | The Inanimate Matter Fields ...................... | 33 |
| Chapter 7 | Animate Matter Fields ................................ | 46 |
| Chapter 8 | Humanity .................................................... | 59 |
| Chapter 9 | Fields of Knowledge ................................... | 68 |
| Chapter 10 | Beyond the Horizon .................................... | 81 |
| Chapter 11 | Design ......................................................... | 85 |
| Chapter 12 | Utilisation ................................................... | 95 |

About the Author .............................................................. 101

# AUTHOR'S NOTES

'Fields' is the author's sixth book and the first book written after the trilogy. The last book of the trilogy 'The Alien World' was published in March 2020.

The author has studied in many of the Fields of Knowledge and wishes to thank his teachers and other authors who may have contributed to the work in the Fields of Knowledge.

'Fields' is linked to the other five books and is a necessary read for the reader who may be interested in any of the other books written by the author. The purpose of this book is to show the importance of the Fields in the lives of both the living and the dead stars of Heaven and Earth. The Fields all around us, appear at least to the author to be essential for our journey from dust to consciousness.

The words in the book might have different meanings for different people. The author is not responsible for any misconceptions that might arise from the words used in the book.

# EXPLANATION OF THE USE OF UPPER CASE

Upper case has sometimes been used to make it easy for the reader to follow the text:

1) A single word might have a vague meaning if considered by itself. However, in the context of Fields, certain single words have a particular meaning and hence are shown in upper case. For example, the word *Space* has a meaning associated with the region where the galaxies exist. Also, *Solar System* is the term used to describe the sun and its surrounding planets.
2) A two-word phrase might be required to be considered together and not considered separately. For example, the phrase *Star System* is associated with a star and its associated planets and moons.
3) The three-word phrase *Fields of Knowledge* must be considered together.

I hope this information is helpful to the reader.

# QUOTE RELEVANT TO THE BOOK

**Thy will be done on Earth as it is in Heaven.**
—The Lord's pray, *The Our Farther*

The words of this prayer are known to most Christians and the words are repeated at every Catholic Mass. Fields are a significant means of how the Creator's will is carried out in Heaven and on the Earth. It is truly remarkable how objects like galaxies can exist for billions of years and have processes that are analogous to life that exists for only tens of years on the Earth. The will of the Creator to bring things into life through birth, then to go through a vigorous youth, middle age, old age and death appears to be universal. Everything experiences the will of the Creator in a similar sort of way.

The words show that whether you are a star in a galaxy in Heaven, or a star in a Field of Knowledge on the Earth, you must go through the same processes of birth, youth, middle age, old age and death. In doing so, you experience and add some spirit to the Fields of the Universe in accordance with the will of the Creator.

# 1
# INTRODUCTION

The human senses are wonderful, constantly sending data to a human brain for processing. To touch, to hear, to smell, to taste and above all to see, enable us to make a sort of picture of the world in our minds.

To our unaided senses, the horizon imposes a limit to the information we can access. Unaided or without our instruments, we cannot see, hear or obtain any information beyond the horizon.

However, today with the accumulated knowledge of generations of humans from the past and billions of humans in the present, we are able to sit at home beside our computers or television monitors and see and hear information from beyond the horizon, in fact from anywhere on the Earth and beyond.

It is truly a miracle how humans achieved such enormous capability in a relatively short space of time.

Time appears to be a very significant parameter in seeing the world, and beyond.

Is the human existence part of an overall design that is part of an overall plan?

## Introduction

The Fields of Knowledge has given us the telescope, the microscope, space probes, radio telescopes and a host of instruments to extend the human senses.

The Fields of Knowledge has shown us the beauty of the macroscopic universe. The stars coming to life within the galaxies. Stars going through the same processes humans experience on Earth. Stars being born, going through middle age, then old age and finally dying. The Fields of Knowledge has shown us the galaxies moving in Space and even interacting with each other over millions of years.

The Fields of Knowledge has shown us the beauty of the microscopic world. Elements being created in the core of stars. Atoms with a central nucleus of protons and neutrons surrounded by negatively charged electrons. The outer shell of the electrons can bond based on the rules of the Chemical Field to form exotic molecules and compounds that can exist as rocks and minerals on the planets and their satellite moons. Atoms given time can form molecules of all kinds on all sorts of worlds. Atoms and molecules can combine if the rules and laws permit, to form life with intelligence and consciousness, as seen on the Earth.

Believing in a Creator and possessing intelligence and consciousness, human beings were able to store data and create the Fields of Knowledge. Through the Fields of Knowledge, human beings are able to see how the dust of the Universe is transformed into elements in the inanimate rocks and minerals found in the Earth's crust. The beauty and symmetry of gemstones and elements can only be observed and appreciated by intelligent minds. The extraction of the elements locked in the rocks and minerals on the Earth is only possible through the technology developed through the Fields of Knowledge. Through the Fields of Knowledge human beings are able to see how animate lifeforms are composed of cells with genes and chromosomes. The manner in which information is stored and passed on to form new cells through mitosis and meiosis is unbelievable. The cells appear to know what they are doing or they appear to be programmed to follow rules and laws even at the microscopic level. The birth of a healthy new born child complete with arms, legs, heart, lungs, kidneys, liver, stomach, eyes, nose, ears, tongue, mouth and intestines all fully functional with repair capability is like a gift from the Universe. The parents did not do any designs themselves, they merely followed rules and provided the sperm and egg cells with chromosomes. The rest was done automatically as the woman's body followed rules to protect and nurture a growing fetus.

Without any intelligence or consciousness the Universe and all its Matter cannot be observed. Without an observer the Universe and all its Matter has no meaning. It is intelligence and consciousness that gives the Universe and all its Matter meaning, and it is intelligence and consciousness that provides us all with a sense of the Creator.

The Universe with its galaxies and stars contains enormous power and energy, time scales in billions of years, distances measured in light years, enormous forces and Fields all of which are easily understood by human intelligence.

The atoms that come to life in a sense on worlds around the stars contain enormous power and energy. They can transform themselves from solids, to liquid, to gases and vice versa. The atoms can automatically use energy from the star to from molecules and chains of molecules on planets and moons. The atoms can convert energy into life through a very interesting coding process that is complex yet accurate, with the ability to maintain complex organ function such as heart beats, digestion, and removal of waste products for at least 70 years and in some cases longer. Amazingly, the life process includes reproduction where information is stored in a single cell that in many cases is born from the fusion of a male sperm cell and a female egg cell. In many human cultures this event is significant, creating bonds of love and friendship. Many religious people believe this to be the work of a Creator. However, many atheist people believe that this remarkable event does not prove the existence of a Creator. Unfortunately, the scope of this book is not to solve this mystery but merely to state the facts about the remarkable Fields all around us.

There is always a tendency for things to go from an ordered state to a disordered state. But this requires time. Just as birth, middle age and old age (disorder) requires time. So, consciousness and mind can appreciate the beauty of Universe and the Creator before everything comes to an end with old age and death.

The stars and galaxies are incapable of writing things down. Human consciousness is special because events can be written down and accumulated through time to gain an appreciation of the Universe and a sense of the Creator.

The Fields of Knowledge help us to see beyond the horizon and to discover new things as time progresses. Through the Fields of Knowledge we can see:

- The existence of the Macroscopic Fields of Heaven namely the Black Hole Field, Galaxy Field, the Star Field, the Planetary field and Moon Fields and the existence of the Microscopic Fields of the atoms and cells.
- The existence of the Inanimate Fields of Matter on Earth.
- The existence of the Animate Fields of Matter on Earth.

As human beings living on the planet Earth, we appear to be surrounded by Fields both in Heaven and on the Earth. It is obvious that the Fields in Heaven and on the Earth are important for our existence. So I decided to look up the meaning of the word 'Field' in my Little Oxford English Dictionary. I found the following meanings obviously dependent on the context in which the word finds itself:

- An enclosure of land for growing crops or keeping animals.
- A piece of land used for a sport or game.
- A subject of study or area of activity.
- An area in which a force has an effect example a magnetic field.
- All the people taking part in a contest or sport.

All these definitions of the word 'Field' are applicable to this book. The reader should note the context in which the word finds itself to discover its true meaning.

One of the lessons in this book is not to prey needlessly on the lifeforms in the biomes, but to pray to the Creator to give us the wisdom to build machines that will help us solve problems on the Earth and enable us to take human consciousness to other stars and galaxies. Perhaps, this is the purpose of the design to have galaxies with Matter and Mind where conscious beings can see the Universe through the Fields of Knowledge.

# 2

# THE BLACK HOLE FIELD

The dust and space was in existence everywhere. A churning ocean of dust in endless Space. Time had no meaning or time meant nothing. There was complete chaos and total randomness.

Then the Fields, Rules and Laws were created to begin the processes that would create the order from the chaos. The order would lead to consciousness, so that the dust could see and understand the Universe.

Some of these Fields are:

2.1   The Gravitational Field
2.2   The Electrical Field
2.3   The Chemical Field
2.4   The Temperature Field
2.5   The Pressure Field
2.6   The Magnetic Field

All the Fields have rules and laws that impose order on the objects within the Field.

When a massive star thousands of times the mass of the Sun is born it burns hydrogen fuel in its core at a very rapid rate. The fusion reactions occurring in the core of the star forms a variety of elements converting hydrogen into heavier elements systematically. The process fuses hydrogen to helium, helium to carbon, carbon to oxygen, oxygen to iron giving out large quantities of energy during all stages of the process. Each stage of the process may last many millions of years. Once iron is produced in the core of a massive star there is a problem, because the rules do not allow iron to fuse to produce the energy required to prevent gravity from pulling the star inwards. The rules and laws do not allow this process of generating energy by fusion to continue forever. The star expands at first and then collapses to the size of a planet sized object. Without the fusion energy available to push outward, the rules force the gravitational energy to pull the mass of the star inwards. Many of the elements formed during the stars lifetime are thrown into the Space around the star and many of the heavy elements are formed during this process. This dramatic event can be seen in time from other galaxies by a brightening then a sudden dimming and disappearance of a star. This event represents the beginning of a new galaxy as the dense and rapidly rotating Black Hole Field is born. The new Field makes the surrounding dust and matter revolve around it. With matter revolving around the tiny massive Black Hole, time begins for the new galaxy. The Black Hole Field is so strong that light energy cannot reach its escape velocity to escape from its surface. Since light cannot escape from its surface, the Black Hole cannot be seen by eyes adapted to receiving visible light. However, stars can be seen revolving around a mysterious point in Space known as the Black Hole Field.

Only the most energetic radiation like X-rays with wavelengths $10^{-8}$ to $10^{-14}$ meters and cosmic rays with even smaller wavelengths of $10^{-14}$ to $10^{-16}$ meters can escape from its surface. The Black Hole Field can usually be detected by scientists observing massive stars revolving around an object that is not visible and generally located in the center of a spinning galaxy.

As the enormous dust clouds containing many of the elements revolves around the Black Hole, the rules and laws of the Universe form the new stars that revolve around the Black Hole. With the birth of a new star time begins for the new Star System. The new Star System and its family of planets, moons, asteroids, meteorites and matter revolve around the central Black Hole. The Fields act on the dust within the new galaxy constantly creating new stars that generate enormous temperatures and pressures in their cores to continue the process of making elements for future stars. The Black Hole continues to rotate forcing the stars and matter to revolve around it. Time continues on for the new galaxy as the stars and matter continue to follow the rules and evolve. The stars go through the process of birth, middle age, old age and death. A process which takes billions of years. In death the stars maintain their mass becoming Dark Matter and influencing the galaxy in which they are located. The dust has not only been converted into stars but into tiny atoms which are distributed in Space by the stars movement and during its death.

The Black Hole Field allows for the creation of generations of stars producing an abundance of elements and unique planets and moons where the elements can follow the rules to exist as exotic molecules and compounds. The Black Hole Field provides the rules that allows the dead stars to remain in the galaxy as Dark Matter which influences the shape of the galaxy. The Black Hole Field also provides the rules that allows the living stars to form and pour out energy by forcing atoms to combine under the tremendous pressures in the center of the star. This energy falls on the planets creating environments with varying temperatures and pressures that allow substances to exist as solid, liquids and gases.

Given a comfortable environment and energy, the atoms can combine in accordance with the rules and laws to form all manner of exotic rocks, minerals and life in the case of a special planet like the Earth. On the Earth we see that life must start out simple. Following the rules and laws, life becomes more complex in a manner that may be described as evolution. It is more like a simple vehicle or aero plane following rules and plans developed by a human mind or collection of human minds and becoming a jet. The process looks automatic but the rules and laws have to be followed for the successful development. If the rules are not followed the process would be corrupt. Also, if the rules and laws were used by corrupt minds the development might be corrupt such as weapons of war and bombs. Because the system is essentially open the rules and laws could be used for corrupt purposes. However, only human minds are capable of large scale destruction and there are generally rules and laws to prevent this.

As life gets more complex and in particular human life there are more rules and laws that are required for control or to prevent the chaos from occurring.

The process from dust in Space to life not only involves Fields, rules and laws, time is an important factor. Time is extremely important for intelligence and consciousness because of the environment and complexity required. The development of the life senses must reach a high level of coordination to produce the kind of life necessary to understand the Universe. However, given the lifetime of the stars it is possible for the dust to turn into intelligent life as seen on the Earth.

The rules and laws result in some dying stars exploding and filling the Universe with new elements. This happens everywhere and must continue to happen everywhere for the process to continue.

The whole process involves enormous quantities of matter and energy. The design involves converting the dust into a more ordered Star System where the star converts mass into energy in its core and produces the energy and Field for its planets and moons. The Inanimate Fields and Animate Fields that can exist in the crust of planets and moons, provide a purpose for the design. Time imposes a limit to the process. We can only have order for a limited time until disorder takes over. Perhaps one of the purposes of the design is for consciousness and the Fields of Knowledge to develop from the Inanimate and Animate Fields to help us to improve our existence on the Earth and also to take our consciousness and the Fields of Knowledge to other stars and galaxies.

The main purpose of the Black Hole Field is to drive Matter around the very dense central core of a collapsed star believed to be a Black Hole. This begins time for the new galaxy. As time moves on the matter continues to revolve around the Black Hole. After millions of revolutions of following the laws and rules, the matter and dust clumps together in many areas forming billions of stars and evolving into the Galaxy Field.

# 3

# THE GALAXY FIELD

The galaxies fill the Universe. There are in estimated 500 billion galaxies in the Universe. A typical galaxy is about 100,000 light years across and 1000 light years thick. The galaxies appear to be a region where living and dead stars exist with their associated planets, moons, debris and dust.

The living stars form in the dust clouds of the Galaxy Field where the rules and laws force matter together until the mass, temperature and pressure is high enough to start nuclear fusion in the core of the massive object. The dimensions and scale of the scheme allows for an Open Star System with live very hot stars undergoing nuclear fusion forming plasma with high temperatures and pressures, while the planets and moons in the Star System can have a variety of temperatures and pressures enabling substances to exist in solid, liquid and gaseous states on their surfaces. The planets and moons can have all manner of landscapes and climates allowing for all types of chemical reactions to take place. When in the Galaxy Field, the planets and moons receive varying energies from their host star keeping them alive and active.

The dead stars appear as Dark Matter. They do not burn nuclear fuel in their cores and have collapsed having high rotational energy and mass they are able to transform and shape the Galaxy Field sometimes producing exotic spiral arms.

The scale of the galaxy, together with the quantity of galaxies indicates a Design that allows for unlimited combinations of mass and matter to form unique Star Systems. To make for even more variety, the galaxies are known to interact with each other pulling each other out of shape. Also, when a galaxies dust and gas clouds are disturbed by an interaction or collision with another galaxy, bright new stars are born leading to spectacular lighting displays for anyone that might possess the senses to see them or consciousness of the Fields of Knowledge. Generally, such events occur over extremely long periods of time and as the stars are far apart there are usually no star collisions.

If there is a purpose to all this, it might be for the dust to become conscious lifeforms capable of sensing the beauty of the Universe. The stars in the Galaxy Field do not seem to be conscious at least in human terms. The stars in the Galaxy Field seem to be following the rules and laws set up or configured into the Fields. This appears to show that the Design of the Universe needs intelligence and consciousness to give it some meaning.

The Galaxy Field varies in appearance, size and shape over relatively long periods of time. Astronomers have fitted the Galaxy Fields into three basic patterns – spirals, elliptical and irregulars.

Spiral Galaxies are easy to identify with their sweeping arms, which contain gas and dust that readily forms into new stars. An important subclass of the spirals are the barred spirals. The barred spirals have a roughly oblong shaped center and may have undergone a collision with a smaller galaxy. Astronomers believe that the Milky Way Galaxy (in which the Sun and its Planets exist) is a Barred Spiral Galaxy.

Elliptical Galaxies contain mostly older stars and little gas to make new stars. A subclass of the elliptical galaxies are the Dwarf Elliptical Galaxies. These galaxies are numerous, however they are difficult to see as they contain dim stars. Astronomers think that these ball or oval shaped galaxies may have formed early in the history of the Universe.

Irregular Galaxies are small and shapeless, but many are still actively making stars. The small Magellanic Cloud is an Irregular Galaxy that is being distorted by the nearby Large Magellanic Cloud and the Milky Way Galaxy.

The Galaxy Field has vast clouds of dust and gas known as nebular. Nebular is the Latin Word for cloud. These regions display spectacular colours for those lifeforms with a sense of vision in the visible part of the electromagnetic spectrum. These are the star forming regions where new stars are born and where the nebular absorbs energy from the new star making its gases glow red – emission nebular. Reflection nebular contain dust grains that absorbs energy from the light of the new star generally glowing blue. Some dark dust nebular are visible because astronomers see them against a bright background. The Horsehead Nebular for example is a big cloud of dusty gas thick enough to block the light from the emission nebular behind it. The above nebular in the Galaxy Field are the birthplace of new stars. Planetary nebular on the other hand form when stars like the sun grow old and throw off their outer layers. The outer layers of the star become a glowing shell of gas that expands into Space and is lit by the hot core of the star.

The main purpose of the Galaxy Field is to form a Field where stars, planets and moons can be born and go through the processes of youth, middle age, old age and death. The Galaxy Field creates Space to house the living and dead stars. The living stars fuse the elements in their cores converting mass into energy, while the dead stars appear as Dark Matter and they shape the galaxy creating the beautiful spiral arms and in some case distorting the galaxy as they evolve and follow the rules and laws that appear to be laid down for them.

# 4

# THE STAR FIELD

Most stars are born in vast dust clouds located in a Galaxy Field. In time, the rules and laws allow all manner of stars to form in the galaxy. Inside nebulae the gravitational fields create dark clumps of matter due to differing rotational speeds around the central Black Holes. If a dark clump of matter is squeezed until its temperature reaches 10 million degrees centigrade hydrogen in the dark clump fuses to helium and the energy released causes the new star to glow. This event starts time for the new Star Field.

    Generally, the same dust cloud that forms the star gives rise to planets, moons, asteroids, comets, meteorites and others debris all revolving in the new field created by the new star. The star and its family of matter, form a Star Field that is very dynamic with matter colliding and increasing the mass of the relatively stable planets and moons. In time, meteor bombardment and collisions get less, until after long periods of time (billions of earth years) one sees a relatively stable star with stable planets and moons, with occasional meteor impacts roughly every one hundred (100) earth years. The Star System is like a basic unit for the galaxy. Typically, two hundred (200) billion Star

Systems make up a typical galaxy. The star revolves around the center of galaxy with its family of planets, moons and associated matter. Stars go through cycles of birth, middle age, old age and death. In old age and death, stars produce the heavy elements which are distributed into the galaxy and are essential for the creation of new stars with planets and moons where the elements may be able to form inanimate and animate lifeforms, if the rules and laws permit it.

The Star Field must have started out as a vast sea of hydrogen. The element hydrogen is the simplest element in the Periodic Table of elements. Hydrogen consists of a single proton with positive charge and an electron with negative charge orbiting around the proton. The sea would have become spherical as the Fields acted on the hydrogen. Eventually, the hydrogen in the central region of the spherical sea began to be squeezed and two protons were pushed together to form the next element helium and release energy. This process of nuclear fusion in the core of a star is the result in a couple of Fields with well-defined rules and laws:

1) There is the electrical field of protons, electrons and neutrons
2) There is a gravitational field pulling the star inward
3) There is the pressure field caused by matter being pulled inward on a very large scale
4) There is the temperature field
5) There is an energy field pushing outward caused by atoms being fused together in the core of the star resulting in a loss of mass which is converted into energy.

To keep the star from collapsing more and more matter has to be fused in the core of the star to create the outward push of heat and energy. Neutrons with neutral charge are formed when electrons are squeezed into the protons. With the start of the fusion process that converts hydrogen into helium, mass is converted into energy and the star is born or time starts for the Star Field. In the young star, the Fields created fuse hydrogen to helium. As time passes usually a few billion years, the supply of hydrogen diminishes, energy must be created by fusing helium to an element with higher atomic number namely carbon. When the helium is diminished, energy must be created by fusing carbon to an element with higher atomic number, namely oxygen. When the supply of carbon is diminished, energy must be created by fusing oxygen to an element with higher atomic number namely iron. The process of fusing elements is extremely long which allows the star to shine for billions of years. However, when iron is created in the core of stars, there is an energy problem, because, the rules of fusion do not allow iron to fuse into an element with higher atomic number. The result is that the star heats up and expands when it fuses elements higher than helium. However, when iron is created in the core, there is no further energy produced because the element iron cannot be fused in the star's core. The star is pulled inwards by the gravitational field causing a massive explosion. The core of the star is compressed and spins rapidly throwing out most of the 118 elements created into the galaxy to create new Star Fields with more exotic elements.

Astronomers work out the size of a star from its brightness and its temperature. The brightness of a star depends on its mass. Mass is generally the size of the dark clump of matter that is brought together in the new gravitational field created by the Black Hole Field. The stars forming with the mass of the Sun represent medium sized stars. The dark clumps of matter can range from 100 times the mass of the Sun to 6% the mass of the Sun.

Stars fall into three basic categories:

1) Large white stars bigger than the Sun
2) Medium sized Stars the size of the Sun
3) Small red dwarves or brown dwarves the size of the planet Jupiter

Large stars are hot and white. The large stars make energy faster and have higher temperatures. They produce fusion reactions that result in heavy elements being formed in the core of the star. Super large stars with a mass 100 times that of the Sun can produce heavy elements including fusing carbon to iron. These stars end up using their fuel and collapsing in a supernova explosion. This results in the formation a Neutron Star typically 20 kilometers across with the mass of the Sun formed from the central core of the collapsed star. The Neutron Star is super dense with a crust of iron and similar elements. The Neutron Stars spin very rapidly beaming out regular pulses of radio waves. They were first discovered in 1960 and called Pulsars because they produce radio waves at regular short intervals typically once every second. X-rays are emitted as material is squeezed on the neutron star by its very strong gravity and powerful magnetic field.

Medium sized stars like the Sun, undergo fusion reactions at a much slower rate than the large stars. Hence they might remain in equilibrium for about seven (7) billion years. Fusion of hydrogen to helium pushes energy out while gravity pulls the star inwards. The star might shine in a steady equilibrium state for approximately seven (7) billion years. When these stars run out of fuel they expand becoming what astronomers call swollen Red Giants. If the Sun were to become a Red Giant its outer layers will probably reach the orbit of Mars. The Red Giant blows of its outer layers and collapses to become an object the size of Neptune with a mass of the Sun known by astronomers as a small dense while dwarf.

Variable Stars are stars that vary in brightness. The variation in brightness is caused by the star flaring up and down. Pulsating Variables are stars that expand and contract. Cepheid Variables are big bright stars that pulsate with energy, flaring up regularly everyone to fifty days. RR Lyrae Variables are yellow super giant stars that flicker and vary in brightness as their fuel runs down.

Binary Stars are double stars that are held together by one another's gravitational field.

Small stars have a relatively slow rate of fusion and are relatively cooler than the other two categories of stars. They are the size of Jupiter and glow with approximately 5% of the Sun's brightness. These stars live for approximately two hundred billion years. They might end their lives forming a crust and glowing very faintly becoming what astronomers call a brown dwarf. Finally, these stars might completely cool of, as the whole star solidifies becoming what astronomers call a black dwarf.

The Gravitational Field that collapses the large dust cloud to form the rapidly rotating star also produces a Star Field with planets, moons, asteroids, comets and debris revolving around the central star. This makes each Star Field unique, each planet unique, each moon unique, and each asteroid and comet unique.

The star provides a new gravitational field for the planets and associated objects, depending on its mass. As the star converts mass into energy, it also provides an energy source for the planets, the strength of which depends on its mass and how it fuses the elements in its core. As a provider of energy to the planets it can affect the climate and temperature on the planet's surface. As a general rule, the planets nearer the star are more effected by the stars heat energy output and are generally rocky worlds, while the outer planets are generally cooler and gaseous worlds.

Thus, the original Field that collapsed the large gas cloud provides the distribution of elements to the planets and moons, determines the positioning of the planets and moons relative the star, provides the asteroids and meteorites that impact the planets and moons and is responsible for the rotational speed of the planets and moons. While the Star Field converts the dust into an energy source and makes the elements by fusing hydrogen in its core. A star is too hot to have matter existing in a solid, liquid or gaseous state. Matter within a star can only exist as a plasma.

The fact that medium sized stars are designed to live in a stable state for about nine (9) billion years, while small stars are designed to live in a stable state for approximately two hundred billion years, implies that the Star Field provides a relatively stable source of energy to act on the elements distributed on planets and moons to transform the elements into a variety of compounds over long periods of time. The energy of the Sun results in various climates on the planets, a typical example is Venus where temperatures are so hot that lead will boil. While on Mars the temperatures are so cold that water is frozen in the desert like soil.

The Design of the System is beyond belief and supernatural because the stars create Fields where the microscopic atoms formed in the cores of stars can form all sorts of interesting compounds that give rise to the Inanimate Fields and Animate Fields on worlds revolving in the Star Field.

Thus, the design or formation of the star is extremely important because it compresses the dust into a more ordered state of a Star Field with the star in the center and planets, moons and matter following the rules and orbiting around the star in time. The laws and rules of the Design force the central core of the star to begin Nuclear Fusion releasing heat energy and electromagnetic radiation in the Star Field for the planets and moons. The rotating planets come complete with raw materials from the gas and dust of the nebular and as they find themselves in differing distances from the star they have differing climates in which the raw materials they possess can develop in accordance with the rules and laws. The whole Design is extremely open, thus increasing variability to the maximum and allowing for greater combinations by making some of the planets with independent Fields to support their moons. This variability of climatic conditions together with the relatively long periods of the stars stable existence makes it possible to sustain Inanimate Fields and Animate Fields where conditions are favourable on a planet or moon.

The stars have their Fields and they follow the set down laws and rules. Stars have life cycles they are born, they are energetic during their youth, they grow old and they die. Unlike human life cycles of 80 years, stars have much longer life cycles generally dependent on their size. There are rules and laws that determine how the star burns its gas based on its size. Large stars burn their gas quicker than smaller stars, hence, the large white stars will remain stable for tens of millions of years, while the medium sized stars will remain stable for ten billion years, while the smaller dwarf stars can remain stable for two hundred (200) billion years. This means that chemical reactions and stable conditions can remain for extremely long periods of time on the small stars.

Also, because the length of the star's stable life is millions of times greater than a human stable lifetime, star deaths appear to be relatively rare in human lifetimes. Also, the night sky appears to be unchanging. This is a design feature that provides stability and security for developing life on a planet and moon.

Essentially, the star provides a Field for its family of planets and moons which revolve around the star and to evolve in time following rules and laws.

The stars existence allows raw materials to exist on planets and moons where they are subjected to varying temperatures and pressures, allowing Inanimate Fields and Animate Fields to develop. The design allows for lifeforms to develop intelligence and mind to understand the Fields, Rules and Laws of the Universe. This could be one of the main purposes of the design because to understand the Fields, Rules and Laws is similar to understanding the Mind of Creator.

The end product of the system being not only to convert the dust into Black Hole Fields, and Star Fields, but to allow for a continuation that allows for the dust to become inanimate matter of elements locked up inside rocks and minerals or animate matter of life and consciousness in the Planetary and Moon Fields.

# 5

# THE PLANETARY AND MOON FIELDS

The Planet and Moon Fields are formed in the Star Field from the same dust cloud that formed the star. As the planet has mass it provides a new gravitational field for its moons and associated objects. The planet Saturn has blocks of ice revolving in a ring system near the planet, while a system of moons revolve around the planet. The rules do not allow planets to undergo nuclear fusion in their cores and so planets provide only a limited amount of energy to their moons. Following the rules, generally moons formed around planets far out from the star are composed of very light substances which are held together in the moons low gravity field. While revolving around a planet, moons can experience tidal forces which can generate heat causing the moons to become active with heat causing volcanic activity. Some substances can reach their melting point and flow as a liquid on the moon's surface. This activity increases the Inanimate Fields and Animate Fields possible on planets and moons of a Star System.

## The Planetary and Moon Fields

Asteroids and comets are known to come into a Star Field. Sometimes these asteroids are pulled into the Planetary Field and they can either crash into the planet adding substances to the planet or they can crash into a planet's moon adding substances to the moon or they may be captured by Planetary Field and become a satellite of the planet. The moons Phobos and Deimos revolving around Mars are believed to be captured asteroids. This keeps the design open allowing for other substances to come into a Planetary Field or Moon Field. Many scientists believe that an asteroid crashed into the Earth 63 million years ago creating a dust cloud that destroyed vegetation and created hard times for the dinosaurs. Many people in the Religious Field will say that the dinosaurs were destroyed by the Creator because they lived aggressive lives and were incapable of developing the Fields of Knowledge. Many atheists would say that the asteroid crashing into the Earth was a random event and the dinosaurs were unlucky to be on the Earth when it happened. The fact is that asteroids and comets certainly make the design less predictable and allow for a more open system with more variety.

The elements created in the core of the star are ejected into Space when the star reaches old age or undergoes a supernova explosion. These elements were trapped in the stars gravitational field because they obviously weighed perhaps a few kilograms because of the enormous mass of the star. When released into Space their weight is reduced because they usually end up in Planetary Fields where the heavenly bodies are far less massive than a star.

On the planets, the relatively mild temperatures and pressures allow elements to exist and combine by sharing electrons in their outer shells. This results in a range of stable compounds. On Earth for example, the temperatures and pressures allow two atoms of hydrogen to readily combine with one atom of oxygen to form the water molecule. In addition there are changes in temperatures and pressures that allow the water molecule to exist as liquid. Removing heat from the liquid transforms the liquid water into solid ice. Adding heat to the liquid water transforms the liquid into a water vapour gas.

So the laws allow the atoms to exist only as a plasma on the surface of a star. While, the laws allow the atoms to exist in a more free state with transformations into solid, liquid and gas on the surface of planets and moons of the Star System. This allows the various elements to combine following the rules of chemistry and to exist on the surface of planets and moons of the Star System. If the design of the Universe requires Inanimate Fields and Animate Fields with life and consciousness to come into existence, then given 500 billion galaxies, each containing approximately 200 billion stars, there is a very good opportunity for various kinds of Inanimate Fields and Animate Fields to come into existence on the trillions of Planetary Fields and Moon Fields within the Universe. This must come about from the away the system is designed and from the laws and rules that control the atoms and stars. The design allows for planets and moons where matter exists in a liquid, solid and gaseous state.

Planets where matter can support a liquid like water on its surface, which can be easily transformed into a gas by the addition of heat from the star and can also be transformed into a solid by the removal of heat, would cause more elements to be readily mixed and seem to have the best chance where the rules and laws of chemistry can bring lifeforms into existence.

From a designers point of view a planet with inanimate matter only, is lifeless and relatively inactive though it may have active volcanos erupting with lava flows heating its surface. From a designer's point of view a planet with animate matter is full of life converting energy into movement and transforming inanimate matter into life. This is very interesting but life requires more rules and controls for its continued existence than inanimate matter which can exist without external continued regular daily nurturing. A design with a life supporting system is far more complex than one without life as can be observed on the planet Earth which is full of complex lifeforms.

The design process for the Universe with the use of Fields, Rules and Laws has resulted in an Open System with dust in Space being converted into the beautiful galaxies with billions of stars providing continuous energy for trillions of planets and moons. The process allows for the dying stars to release elements into the galaxy and to produce the Dark Matter that is capable of shaping the galaxy to form new stars which will ultimately form new planets and moons.

The design allows for the planets and moons to have a range of 118 elements together with varying temperatures and pressures which allows for the creation of inanimate matter that can exist as solids, liquids and gases and if more rules are followed even animate matter can form.

The elements are substances that cannot be broken down into simpler substances by chemical methods. There are 94 naturally occurring elements and 24 synthetic elements.

Elements are classified as metals and non-metals.

Metals are shiny solids that conduct electricity. Most metals melt at high temperatures. Metals are malleable, which means they can be hammered into different shapes. Most metals are also ductile, which means they can be stretched without breaking.

Most non-metals melt at lower temperatures than metals, and many are gaseous at room temperatures.

Generally, elements can combine with other elements to form compounds. Chemical reactions can also be used to break down compounds and free the elements they contain.

The synthetic elements are formed on Earth in nuclear reactions using natural elements. The synthetic elements are so unstable that they decay and fall apart, often in minutes or even fractions of seconds.

Uranium is the heaviest naturally occurring element. It has 92 protons, 92 electrons and 146 neutrons. Uranium is formed during some sort of stellar supernova explosion. Elements heavier than uranium are unstable. Their nuclei burst apart because the forces that draw protons together are not strong enough to overcome the repulsion between their positive charges.

The 118 known elements form a pattern when arranged in increasing atomic number. Fortunately, human beings have worked out the pattern and they make a very interesting table called the Periodic Table. On this table, elements arranged in order of increasing atomic number, with similar elements grouped together. The Periodic Table lists the 118 elements formed the core of stars and distributed to the planets and moons, in eighteen vertical columns or groups and seven horizontal rows or periods. The elements are arranged so that their atomic number increases from left to right through the horizontal rows. An element's atomic number is the number of protons in its nucleus and the number of electrons orbiting the nucleus. Elements in the same group have similar properties. The chemical properties of an element depend largely on the number of electrons in the outermost shell of the element. The electrons are arranged in shells around the nucleus.

On the planets and moons, given the right conditions, the Rules and Laws allow element interaction, dependent on the electrons in the outermost shells only. This allows for the formation of many very interesting compounds which can exist in a stable form on the planet's surface or the moon's surface.

As an example, consider the water molecule, hydrogen oxide ($H_2O$), which consists of two atoms of the element hydrogen combined with one atom of the element oxygen. Oxygen has atomic number 8, as it consists of two electrons in its first shell and 6 electrons in its second shell. That leaves space for two electrons to fill its second shell, which can hold a maximum of eight (8) electrons. Hydrogen has atomic number 1, as there is only one electron in its outer shell. That leaves space for one electron to fill its outer shell, which can hold a maximum of two electrons. Hence, two hydrogen atoms can combine with one oxygen atom to form a water molecule. The rules allow this combination to occur. Obviously, if conditions are satisfactory, and if there is plenty of oxygen and hydrogen available on a planet's surface, water will form in abundance. In addition water is a polar molecule because the oxygen has a slight negative charge while the hydrogen has a slight positive charge. These charges attract water molecules to each other and makes water an excellent solvent.

The other factor is the state of matter which is dependent on the planet's energy from the star and its internal energy from its own internal core. Matter on a planet or a moon's surface can exist in three basic forms, namely, solid, liquid or gas. On the Earth for example the average temperature is 25° Centigrade. At this temperature water exists as a liquid. Hence, water gathers in low lying seas and oceans. At the poles the temperature drops to 0° C and heat energy is removed from the water converting the liquid water into solid water or ice. In the equatorial regions the sun heats the liquid water, the added energy causes large quantities of the surface water to evaporate forming clouds which drift towards the land. When striking a mountain or raised land, the clouds of water vapour rise. The resulting cooler condition removes heat energy from the water vapor gas converting it to a liquid and it falls as rain. The rain dissolves minerals and salts out of the rocks on the mountain and carries them via rivers to the seas and oceans. Even the solid ice is known to flow over the land as a glacier carrying rocks and minerals around with it. This ability of water to move around the land powered by the energy from the Sun is crucial in bringing all sorts of minerals and elements into the oceans of water. 70% of the Earth's surface is covered with water. The energy from the Sun keeps the liquid constantly moving and full of energy that is stored as heat within the water. Water also exerts pressure on the thin ocean crust getting elements and chemicals directly from the Earth's internal core. All these factors make it ideal for simple lifeforms to emerge from the energy in the water following the rules that allow the elements to bond with each other and form compounds.

In any substance, particles are in constant motion. This energy called the kinetic energy, increases with temperature. Whether a substance is a solid, liquid or gas depends on the balance between the kinetic energy and the forces of attraction due the electrical charges binding the particles.

Substances are solids when the forces of attraction between the particles are strong enough to prevent the particles from moving freely. Solids have fixed shapes called lattices.

Liquids are fluid and can change their shape. They have flat surfaces and collect at the bottom of a container. In a liquid the forces of attraction between particles are too weak to hold them in a rigid formation. The particles glide past each other.

Substances exist as gases when the kinetic energy of their particles is large enough to completely overcome the forces that attract them. Gases like liquids can change their shape to fit their containers. Unlike liquids, however gases have enough kinetic energy to spread out and completely fill their containers.

The melting point of a substance is the temperature at which the kinetic energy of the substance's particles is just great enough to free them from the rigid lattice solid structure. The amount of energy needed to melt a solid depends on the strength of the attractive forces in the solid.

The freezing point of a substance is the temperature at which the kinetic energy of the substance's particles is reduced to the point that the substance forms a rigid structure. For example liquid water freezes to form ice at 32° Fahrenheit.

The forces in iron, which melts at 2,765° Fahrenheit are much greater than the forces in ice, which forms at 32° Fahrenheit.

A liquid boils when bubbles of vapor that grow in the liquid, rise to the surface and burst. The boiling point of a substance is the temperature at which the kinetic energy of the particles of that substance are great enough for them to completely escape the forces that pull the particles together. Each substance has its own boiling point. Water boils at 212° Fahrenheit to form steam. Liquid hydrogen boils at -436° Fahrenheit and ethanol boils at 174° Fahrenheit. Some substances turn into a gas without passing through a liquid stage. This process is sublimation. Solid carbon dioxide sublimes, it becomes a gas at -173° Fahrenheit.

The above mentioned temperatures are based on sea level pressures of one atmosphere. Temperatures vary for higher pressures.

There are rules and regulations regarding how the elements form bonds based on the outer shell containing electrons. Some of these rules are as follows:

1) Each atom of an element contains a number of electrons that exactly equal the number of protons in its nucleus. The positive charges of the protons balance the negative charges of the

electrons, and the atom has no overall charge. The electrons orbit the nucleus in layers called shells. There is a limit to the number of electrons that each shell can hold. The first shell, which is closest to the nucleus, can hold up to two electrons. The second shell can hold eight electrons, and the third shell can hold eighteen.

2) The rows of the Periodic Table list elements in order of increasing atomic number, which is the number of protons in the atom of an element. Each row starts with an element that has only one electron in its outermost shell. At the end of each row is a noble gas, which has a full set of electrons in its outermost shell. The latter arrangement of electrons, which is called a configuration, is usually stable. Hence, noble gases seldom react because their outer shells have a full set of electrons. Elements react when bonds form between atoms as they gain, lose or share electrons. As a result of these changes, each atom in a compound usually has a full outer shell of electrons. This is the stable electron configuration of the noble gas closest to each element in atomic number. The valency of an element is the number of bonds it must make to attain a noble gas configuration.

3) Metals usually have one or two electrons in their outer shells. They easily lose these electrons so that the next shell down becomes a complete outer shell. The non-metals at the far right of the periodic table are only one or two electrons short of a complete shell. Hence, they easily accept electrons from atoms of other elements. The valences of these elements are the number of electrons they must gain or lose to form a complete outer shell.

4) Elements in the middle of the main block of the Periodic Table have outer shells that are three or four electrons short of a full shell. Carbon is an example of an element in the middle of the main block of the Periodic Table. Carbon has four electrons in its outer shell, which can contain a maximum of eight electrons. Carbon rarely accepts four electrons, because

the negative charge of the ten total electrons would repel the electrons from the protons in the nucleus, making it hard to form a compound. Instead, carbon atoms overlap their outer shells with shells of other atoms and share four electrons to make up the full count. The valency of carbon is 4.

5) Ionic compounds form when atoms of two or more elements trade electrons to form charged particles, or ions, of each element. The ions have a noble gas configuration and their charges balance each other to give the compound no overall charge. For example, Sodium Chloride where Sodium loses its eleventh electron to become a positive charged sodium ion and at the same time chlorine gains an electron to become a chloride ion. The opposite charges of these two ions attract each other strongly. The ions bond together in a regular pattern called a crystal lattice. In a similar manner, the magnesium atom loses two electrons and two chlorine atoms gain an electron each to form Magnesium Chloride. Magnesium has a valency of 2. Ionic compounds form mainly between metals at the left of the periodic table and non-metals at the right of the periodic table.

6) Covalent bonding occurs when non-metal elements share outer shell electrons to make up a complete shell. For example in carbon dioxide, the valency of carbon is 4 and oxygen is 2. In Carbon Dioxide, one Carbon atom shares one pair of electrons with each of two oxygen atoms. In this way, all three atoms fill their outer shells.

Hence, it can be seen that many of the 118 elements listed in the Periodic Table are atoms with outer-shell electrons that can join together to form many millions of different compounds with different properties. Strong forces of attraction called chemical bonds hold atoms together in these compounds.

Following the rules of the Chemical Field the elements on planets and moons can be transformed into all manner of compounds. Given that each galaxy contains billions of stars, means that each galaxy has hundreds of billions of planets and moons where different conditions can arise to create all manner of compounds. Fortunately, here on Earth the conditions are right for the dust to form seas and oceans with millions of chemical compounds contained in the liquid water. This has made it relatively easy for lifeforms to be created and maintained.

Heat and pressure are known to increase the rate of chemical reactions. Heat energy makes the atoms vibrate rapidly making them reactive while pressure forces the atoms together making them more reactive. Both heat energy and pressure are present in the seas and oceans of Earth, together, with currents and tides. The elements would be forced to combine in every way possible resulting in simple plants that could absorb energy and make food to support themselves in the water environment. These plants would have simple inputs, taking in Carbon Dioxide and sunlight to produce internal energy for growth and development releasing simple outputs like Oxygen. From the water, these plants would then spread to the land surface where water was always present underground and in rivers flowing across the land. Over millions of years carbon atoms are lost and stored in plants living and dead, while oxygen is put into Earth's atmosphere. Then, the chemical compounds form animal creatures to live of the plants and control their abnormal spreading. These animal creatures leave the crowded polluted seas to come onto land where plants thrive. Ecosystems and biomes form where creatures live in communities to survive. The complex nature of lifeforms that range from single cells to multicellular organisms with brains, senses, limbs, hearts, kidneys, and lungs show a complex design. We are fortunate to be given this design free by the Universe and the Creator.

The material or matter in each of the planets or moons in the Planetary Fields are subjected to all manner of forces as the planet or moon evolves from a hot ball of gas and plasma to a cooler sphere of plasma with a crust. The elements are squeezed, compressed shaped and reshaped during the life time of a planet or moon. Like everything else the planets and moons follow the rules and go through the process of birth, youth, middle age and old age. The planets and moons are born as molten balls of gas and liquid. In their youth they might have intense volcanic activity as the heat and gas escapes from their molten cores.

Then, they cool down in middle age forming a crust with relatively little or no volcanic activity, as there is less heat and gas available to escape from their cores. As time progresses the elements on the surface are locked up in compounds that are visible as rocks and minerals. The elements found in rocks are mixed with all manner of other substances due to the enormous forces and pressures that brought them to the planet or moon's surface. Usually the elements like gold, silver and iron processed from enormous quantities of rock are known as ore. The minerals of no value are known as gangue. Human beings have not only found the elements of gold, silver and iron but other precious gems like diamonds, sapphires, rubies and emeralds. The existence of such precious and beautiful objects only accessible to human beings could be seen as the reason for a Creator. It is unimaginable to believe that these beautiful objects could exist without anyone ever knowing of their existence.

Of all the planets and moons in the Solar System, there is evidence that lifeforms exist in the biosphere of the Earth's crust. However, the Fields of Knowledge have been in existence for a relatively short period of time and it is therefore impossible to tell what one million years will reveal about the Creator's design. There is scientific evidence that Earth came into existence 4.5 billion years ago and that life existed in earth's crust approximately 3 billion years ago. All life requires food, water, air and energy like sunlight or the Earth's internal heat in order to exist. All these ingredients are provided in the biosphere of the Earth. All these ingredients are recycled, accept sunlight which must be provided constantly by the Sun and the Earth's internal heat. The processes of life follow rules and laws as the lifeforms are born, go through youth, middle age, old age and death. This gives lifeforms an ability to see, hear, touch and experience the creation in many ways. The existence of life may be seen as a reason for a Creator. It is difficult to believe that life would come into existence and obviously remain in existence to observe not only the inanimate world but the animate world for no definite purpose.

Essentially the planet provides a Field for its family of moons which revolve around the planet and evolve in time. A planet produces far less energy than a star and it has far less mass, hence the Planet Fields are smaller in size and the moons are not dense or have relatively smaller mass. Moons that are generally round in shape must have formed from the initial gas cloud that formed the Star System and have molten cores but they would all be composed of lighter substances. Many moons revolving around planets are asteroids and debris captured by the Planetary Field.

The rules and laws for moons travelling around a planet allow heat to develop on the moon. A moon rotating and revolving relatively close to a planet can cause its surface facing the planet to be pulled up and as it rotates away from the planet its surface is pulled down. Since it is composed of very light material this could cause massive tidal forces and heating on its surface. In other cases fairly stable conditions can exist with mild climate because the planet does not supply much heat to their moons. Tidal effects can cause liquids to form relatively easily. Liquids can move relatively easily on a moon's surface and mountains can also form relatively easily in the lower gravity, allowing for substantial activity.

Generally, moons moving around planets can be relatively active and not dead worlds. The rules and laws described for the planets and in particular the chemical reaction and combining of elements into compounds apply to the moons. The rules and laws ensure that the planet provides protection for the moons, because it has a much greater mass it attracts debris and meteorites which would otherwise crash into the moons revolving in its Field. This generally provides more stability for the moons revolving in the Planets Field.

Hence, the Moon Field is capable of allowing the rules and laws that support the Inanimate Field and the Animate Field.

At any rate, we must all be grateful and thankful to the Universe and the Creator because the cost of producing our star and the planets and moons would be astronomically more than all the jewels and gold available on Earth and all the other planetary bodies and moons in our tiny Star System. So we would never be able to repay the Universe or the Creator for the creation given so freely to us.

# 6

# THE INANIMATE MATTER FIELDS

The planets in the Star Field and the moons in the Planetary Field are located at different distances from the star and planet respectively. When the planets and moons form they are in a molten state. As their mass is insufficient to fuse elements into the heavier elements they do not produce energy by fusion like a star. The planets and moons once formed begin to cool from a hot molten state and form a crust of rocks and matter on their surface, with a molten and solid inner core. The crust of a planet and moon is the perfect place for Inanimate Matter Fields to develop. The Solar System is a typical Star System where different kind of Inanimate Matter Fields developed in its planetary field.

The main portion of the dust cloud condensed to form the Sun while lesser portions of the rapidly rotating dust cloud collapsed to form the planets and moons. The collapsing dust cloud resulted in the Sun, planets and moons possessing spin. While the Sun possessed enough mass to allow it to start nuclear fusion in its core and hence emit electromagnetic radiation, the planets and moons revolving around the Sun became molten with the intense heat from the process of the gas cloud (energy) becoming mass. Over time the planets and their moons would cool down from the rules of thermodynamics, usually forming crusts of rocks on their surfaces while their cores usually remained in a molten state. The process of the gas cloud reducing in size to become a molten sphere results in the rotation of the planet and moon. The rotation of planets and moons cause variations in temperatures and pressures that have profound effects on the formation of inanimate matter, resulting in a variety of rocks and minerals.

Today we see a Solar System with Sun that is 4.5 billion years old. The Planet Mercury has a radius of 2440 kilometers and revolves around the Sun in 88 days. The Planet Venus has a radius of 6052 kilometers and revolves around the Sun in 224 days. The planet Earth has a radius of 6378 kilometers and revolves around the Sun in 365 days. The Planet Mars has a radius of 3396 kilometers and orbits around the Sun in 687 days. The Planet Jupiter has a radius of 71492 kilometers and orbits around the Sun in 4333 days. The Planet Saturn has a radius of 60268 kilometers and orbits around the Sun in 10759 days. The Planet Uranus has a radius of 25559 kilometers and orbits around the Sun in 30685 days. The Planet Neptune has a radius of 24764 kilometers and orbits around the Sun in 60189 days. This rule of planets forming and rotating and revolving about the star gives the planets properties. Also, the rule of moons forming and rotating and revolving about a planet gives the moons properties. The planets and moons begin to have atmospheres, molten cores and a crust based on the substances they possessed when the great gas cloud collapsed. The planets and moons have spherical shapes according to rules where matter is pulled towards the center. The heat and pressure in the core tends to drag the heavy elements like iron into the center of the sphere while the lighter elements form the surface crust of rocks.

In this way the rules formed the present Earth with the following properties:

Crust of water and land 30 kilometers thick on average.

Surface Pressure 1.014 millibar.
Average Temperature 15° Centigrade.
Temperature Range -90° to 50° Centigrade.
Wind Speed 0 to 10 meters per second.
Crust of two types relatively thin oceanic crust and relatively thick continental crust.

The crust of the Earth is like the skin of an apple. Under the crust is a deep layer of hot, soft rock called the mantle. Beneath the mantle is a hot iron and nickel core. The inner core is 7000° C and stays solid because the pressure is 6000 times greater than on the surface. The outer core is 4500° to 6000° C and is always molten. The Oceanic Crust is relatively thin approximately 7 kilometers while the continental crust is relatively thick approximately 80 kilometers.

Obviously, from the rules of thermodynamics, the molten rock from the mantle wants to escape to the surface to form rocks and minerals in the surface crust.

Igneous rock is made when hot molten magma or lava rises to the crust and cools and solidifies. Basalt is an igneous rock that forms from lava that has erupted from a volcano (a volcanic rock). Granite is an igneous rock that forms when magma solidifies underground (a Plutonic rock).

While volcanoes are an example of heat escaping from Earth's internal core, earthquakes and tsunamis are an example of gravity pulling the crust inwards. Stable regions are where the outward push and inward pull are balanced. Whenever a volcano erupts or an earthquake occurs, the area is restored to a stable state – a very significant rule. However, damage to the crust is a problem for lifeforms in the region.

The relatively thin ocean crust makes it easy for lava to come to the surface and cause the ocean floor to spread. The ocean tends to put pressure on the internal magma and again the ocean floor expands until equilibrium is restored. This force is significant because it can push the land surface above around the oceans, creating mountains and valleys.

## The Inanimate Matter Fields

In addition to the above the heat from the Sun evaporates vast amounts of water from the oceans which rises into the atmosphere and eventually falls to the ground as rain. This rule causes rivers and streams to form as gravity pulls the water to the lowest points on the land, namely the lakes and oceans. This process along with the wind and ice cause erosion and weathering of the crust. Sediments from the erosion and weathering process are dragged by water, wind and gravity into the lakes and oceans. The result of this process is the buildup of sediments into layers or strata. Hence, sedimentary rocks are made from the slow hardening of sediments into layers or strata. Sandstone is formed on the seabed from sand, silt and rock fragments that are broken down by weathering and erosion processes. Limestone and Chalk are sedimentary rocks made mainly from the remains of sea creatures.

Metamorphic Rocks such as limestone forms when other rocks are changed by extreme heat and pressure. Contact with hot magma turns limestone into marble.

The Earth's crust is very dynamic but the changes occur over relatively long periods of time. Hence, it appears static for a human generation and could only be seen when recorded over generations. The crust is being pulled inward by gravity and pushed outward by the heat energy from Earth's internal core. The land moves as the ocean floor spreads. Where the land converges mountains form and where the land diverges new sea floor is created. Flowing water from rain and ice is constantly eroding and weathering the crust recycling the rocks and constantly moving elements around to make it relatively easy for lifeforms to exist.

In this way, Earth becomes a relatively cool place where water is distributed as rain to keep the land moist making it relatively easy for plants to obtain nutrients and sunlight. The Sun heats the water and creates an atmosphere of water vapour which falls as rain or snow. The flowing water and ice create rivers and streams as they flow into lakes and oceans. The rivers and streams also carry elements which allow chemical reactions to take place in the lakes and oceans. The Sun provides enormous quantities of energy daily on the Earth. The energy powers the climate creating regions of high pressure and low pressure. The Water Cycle moves water around the planet while the Rock Cycle moves materials and elements around the planet.

Minerals are mixtures of naturally occurring chemicals, or elements listed in the Periodic Table of elements. Minerals grow in crystals found in nature and not made from living things such as wood.

Mineral crystals have flat faces and straight edges and grow in six basic shapes:

- Cubic crystals like pyrite crystals which have six square faces.
- Tetragonal crystals like zircon crystals which are cuboid or a stretched out cube.
- Monoclinic crystals like the gypsum crystal which looks like a squashed rectangle.
- Hexagonal crystals like quartz crystal.Orthorhombic crystals like the topaz crystal.
- Triclinic crystals like the axinite crystal.

Rocks are a mixture of minerals and the remains of living things such as shells. There are three types of rock:

- Igneous rocks form when hot molten magma coming from deep within the Earth cools down. This occurs around volcanoes. Igneous rocks are made of a mosaic of mineral crystals, usually without layers. Intrusive igneous rocks form underground in large masses called batholiths and in relatively small intrusions known as sills and dykes. Extrusive igneous rocks form in the cones of volcanoes.
- Sedimentary rocks are made when small pieces of other rocks called sediments are gathered together by natural process of erosion, weathering, water flow, and wind. These sediments

are buried deep within the Earth over long periods of time. The pressure converts the sediments into a rock.
- Metamorphic rocks are made when rocks are melted and squeezed by the pressure and temperature existing on a planet. The rocks are squashed and heated until they form new rocks.

Examples of rocks are:

Granite forms when magma cools slowly, deep within the Earth. It contains coarse grained crystals generally 5 millimeters across which can be easily seen with the naked eye. It is composed of light coloured minerals such as feldspar, quartz, and mica. Granite is easily weathered and decomposes to form sand and clay.

Granodiorite is a common form of granite with the main minerals being feldspar, quartz, hornblende, augite, and mica.

Gabbro forms when large masses of magma cool slowly. Gabbro is found in thick sheets of igneous rock. It contains the two main minerals of feldspar and pyroxene and less than 10% quartz. Gabbro is heavier than granite because it contains a large amount of the heavy mineral pyroxene.

Pegmatite is an igneous rock that forms deep underground and is made up of large crystals 3 centimeters to 1 meter long. Pegmatite crystallises from magma and other high temperature fluids, which are rich in rare elements like niobium, tantalum lithium and tungsten.

Porphyry rocks are found in minor intrusions of magma that emerge from the main batholith of magma. The minor intrusions are known as sills and dykes. The rocks are composed of medium grain size and contain large crystals set into a finer ground mass.

Xenoliths are rocks found around the margins of many igneous intrusions, where magma has melted and forced its way into other rocks.

Syenite is an intrusive igneous rock that is formed by the cooling of magma deep in the Earth's crust. Due to the slow cooling associated with high temperatures at great depth, syenite has large crystals and is a coarse-grained rock.

Dolerite is a dark-coloured igneous rock, often with an overall speckled appearance. This rock is dense and heavy mainly because it contains minerals rich in iron.

Serpentinite is formed from the chemical alteration of other igneous rocks. It contains attractive colours such as shades of greens and reds. These colours appear as veins through the rock.

Obsidian forms when volcanoes erupt explosively, the magma meets the air and water quickly and freezes. It cools so fast that mineral crystals and the matter becomes a type of glass.

Basalt forms when red hot lava flows. It forms on the Earth's surface and on the ocean floor. It is a fine grained rock. When it erupts basalt lava contains much gas. As the rock cools, gas bubble hollows are left in the solid rock giving it a rough texture. The dense rock contains iron rich minerals.

Andesite forms when solidified lava erupts from violent volcanoes. The devastating eruption of Krakatoa in 1883 produced millions of tons of andesite lava.

Rhyolite forms when violent volcanoes erupt producing lava with a large percentage of silica. This makes the lava viscous or sticky. As the sticky lava flows down the volcano it blocks vents as it cools, thus building up pressure. Some acidic lava volcanoes develop a tall spine of solidified rhyolite lava above the cone.

Unakite forms from granite and a green mineral called epidote. It has a colourful speckled appearance.

Pumice forms in volcanic eruptions where there is a lot of gas in the molten lava. It is a very light rock containing many holes and pits on its surface.

Diorite forms when there are light and dark minerals in the lava.

Limestone forms from the shells and skeletons of ancient ocean dwellers, such as shellfish and corals. It contains a high proportion of the mineral calcite or calcium carbonate. It is a sedimentary rock.

Flint forms from the mineral quartz and is composed of silicon dioxide. It is a sharp rock that can be broken into shards. In ancient times it was used to make arrowheads and spears. Flint is found in irregular layers of chalk. The silica in flint is derived from small sea creatures.

Sandstone forms when a layer of sand is compressed perhaps under the ocean floor for thousands of years. The sand sized grains form a rock. The main mineral in sandstone is quartz.

Chalk is a special type of limestone that is almost pure white. It is very fine grained and contains more than 90% calcium carbonate. All the calcium carbonate in chalk are the dead remains of microscopic and small sea creatures that lived in or on seabed millions of years ago.

Shale forms from soft minerals such as clays. It is a common type of sedimentary rock found below the soil.

Coal forms from the fossilized remains of plants that lived long ago in swamps or bogs. The deeper the coal is buried, the denser it is and the more energy it contains.

Marble forms from contact metamorphism of limestone. Heat from an igneous intrusion or lava flow causes the calcite in the limestone to recrystallize. Marble is made up mainly of the mineral calcite and contains veins of other types of minerals.

Metaquartzite forms from contact metamorphism of sandstone. Heat from an igneous intrusion or lava flow causes the quartz grains in the sandstone to recrystallize. Metaquartzite is crystalline, non-porous very hard rock.

Slate forms when sedimentary shale or mudstone experience high heat and pressure becoming a relatively strong and hard metamorphic rock.

Schist forms if slate is buried deep and subjected to heat and pressure. New minerals are created that turn the sedimentary slate into metamorphic schist. The minerals form layers stacked one on top of another known as foliation.

Lapis Lazuli forms from the mineral lazurite which gives it its characteristic blue colour. Other minerals include sodalite and gold flecks of pyrite.

Gneiss forms under extremely high temperatures and pressures. The layers within the rock are squashed into folds by the high pressure which results in bands of grey, pink or white rock. The heat and pressure in the process of mountain building turn mudstone into slate, then schist and finally into gneiss.

Hornfels forms when fine grains of mudstone are baked by a nearby source of heat and not subjected to intense pressure. Hornfels is hard and strong like a brick baked in a kiln. It is a metamorphic rock.

Eclogite forms under high temperature and very high pressure in the deep roots of mountain chains. It contains large coloured crystals.

Examples of minerals are:

Quartz is one of the simplest and most common minerals on the Earth. It comes in a range of colours.

Topaz is much harder than quartz.

Olivine crystallises at very high temperatures in basalt rocks. It is a silicate of iron and magnesium. It has small greenish grains or crystals that appear brown when it has a high iron content. Refer to peridot below.

Corundum is an aluminum oxide that forms in igneous and metamorphic rocks. It is the second hardest mineral to diamond. Refer to rubies and sapphires below.

Beryl forms in igneous rocks such as granite and pegmatites. Refer to emerald and aquamarine below.

Amazonite is a blue-green variety of the mineral microcline.

Haematite is an iron oxide, while bauxite is an aluminum ore.

Pyrite is made up of iron and sulphur. It has cube shaped crystals and a gold metallic lustre.

Jade contains to two minerals jadeite which has a pale green colour or nephrite which is slightly softer and has a glassy or greasy lustre.

Tourmaline is found near granite, it is a very brittle mineral.

Mica is made of layers of many flat, sheet-like crystals.

Moonstone is a mineral that got its name because it reflects light to produce the effect of moonlight dancing on water. It contains ultrafine layers of a mineral called adularia that reflects light.

Chalcedony has very tiny crystals which can be seen only with special equipment. Hence, they are smooth and glassy.

Geode forms when water containing dissolved minerals seeps into hollow spaces in a rock. The water disappears due to heat and the dissolved crystals are left behind inside the rock. They are found inside basalt or limestone.

Garnet is a reddish gemstone. It is hard and can be used to grind down softer gems.

Amethyst is a purple colour gemstone. It is a variety of quartz.

Aquamarine is a blue variety of the mineral beryl.

Diamond is a form of the element carbon. It forms small, glassy, often octahedral crystals in the cubic crystal system and is the hardest known mineral. The atoms of diamond are joined in groups of five. These link together in a tight, close structure.

Graphite is a form of the element carbon. Graphite is very soft and can be scratched with a fingernail. Graphite is used in pencils for writing. In graphite the atoms are arranged in layers or sheets that are weakly joined.

Emerald is a gemstone with a deep green variety of the mineral beryl.

Ruby is a gemstone with red crystals of the mineral corundum.

Peridot is a gemstone with green crystals of the mineral olivine.

Sapphire is a gemstone with blue crystals of the mineral corundum.

Labradorite has iridescence in light. The mineral crystals split the light into a rainbow of colours.

Sodalite is a deep-blue coloured mineral that contains the metal sodium. It is light and breaks easily. It is one of the minerals inside the rock lapis lazuli.

Turquoise is a gemstone with a soft bluish or greenish colour.

Fluorite glows under ultraviolet light. It grows in cubic crystals and has a blue to purple colour.

Rhodonite is a gemstone with pink or rose-red crystals. It is an ore of the metal manganese which gives it its pink colour.

Opal is a gemstone that produces flashes of colour because of the packing of minute spheres of silicon dioxide and water in its structure.

Some gemstones are distant relatives of ancient trees and animals like amber, oyster, petrified wood, pearls and shells.

Elements and minerals are found in rocks where they are usually twisted and distorted due to the temperatures they have encountered in coming into the crust of the planet or moon and by the movement of the crust over long periods of time. Usually the elements and minerals have to be extracted from the surface rocks by mining, then they have to be processed and separated from the invaluable surface rock.

An ore is a piece of rock that contains sufficient of an element or mineral to make it worthwhile mining and processing.

Magma chambers are large bodies of molten rock that form within the Earth's crust. The upper parts of the magma may have up to several per cent of water dissolved in them. When the magma chamber ceases to become active and begins to cool, the molten magma begins to solidify by creating crystals. During crystalisation, the first minerals to form are those that do not incorporate much water like feldspar. This increases the water content in the remaining magma. The material is more runny and moves over a wider area creating larger crystals called pegmatites. Pegmatites are a source of many rare metals, sheets of mica and some gemstones.

Many closed spaces are filled with minerals that have formed from hydrothermal solutions. When the hot solutions cool the water evaporates out, leaving veins of minerals and elements in the rock. Metals like gold and silver are formed in veins within rocks in this manner.

Generally, the cooling magma follows many different rules as it reaches the surface and cools. Earths huge quantities of water assist the cooling processing creating the variety of minerals and gemstones described above.

All the elements and gemstones have to be mined, processed and specially cut to reveal their beauty. Only beings with eyes and brains can appreciate the beauty of the Inanimate Matter all around us through the Fields of Knowledge.

The formation of the beautiful structures like gemstones from inanimate matter is like a miracle. The process is driven by intense heat and energy. However the end products when refined are an indication that there is more to the process of converting dust to beautiful objects on a planet's surface. It is difficult to believe that the process is purely random, it appears to have some creative force or design embedded in it.

There are two types of common rocks that are not directly related to the original gas cloud that formed the molten Earth. These rocks are coal and petroleum.

Coal is a metamorphic rock formed from ancient plants that were buried for long periods of time and subjected to intense heat and pressure. Metamorphism of organic matter needs a far lower temperature and pressure than inorganic rocks which are obviously harder substances. As a result coal is found interbedded with sedimentary rocks. The degree of metamorphism determines the amount of carbon in the rocks and hence the purity or energy the rocks can release.

Coal that forms in thick layers of poorly decaying plant material in swamps and bogs, forms low carbon and low energy peat. Sometimes the plant matter is covered over by a layer of sediment that prevents the air getting to it and oxidising the peat to carbon dioxide gas. This requires the land to sink compressing the peat and removing a large portion of the water content. The peat becomes transformed into a soft brown coal known as lignite. With further metamorphism and the removal of more water the lignite is transformed into a harder coal known as bituminous coal. The driest coal with the highest carbon content and the highest metamorphism is known as anthracite.

The other rock petroleum comes from the two Latin words petra and oleum meaning rock oil. This substance is formed from the decaying bodies of microscopic sea life and is found in sedimentary rocks. Petroleum includes oil and natural gas. Petroleum is a mixture of oil and natural gas. Petroleum is a mixture of hydrocarbons and the decay of the microscopic organisms is a major factor in forming this rock.

Both coal and petroleum are different from other rocks because the carbon stored in the rocks are an important and abundant source of fuel. These rocks are yin-yang rocks, because they have a yin side when used as a source of energy to power trains, ships and cars making it possible for the Industrial Revolution and easy energy available for powering machines. They also have a yang side because in creating energy they release the carbon which had been stored up in these rocks over millions of years. Carbon is a greenhouse gas which when removed from the atmosphere makes the Earth a cooler place. Many scientists believe that by releasing large amounts of carbon into the atmosphere it is possible to have global warming resulting in climate change. It indicates that changes to natural processes could result in long term effects which might not be fully realized. It also shows how these inanimate rocks could mysteriously exert a controlling effect on Earth's atmosphere to keep the temperature and climate cooler.

# 7
# ANIMATE MATTER FIELDS

Inanimate matter in the form of rocks can exist for billions of years while animate matter can exist for tens of years as individuals and many hundreds of years as a species. Inanimate Matter is static while Animate Matter is progressive and with the proper Fields, Rules and Laws can lead to intelligence and consciousness. Intelligence and consciousness with the proper Fields, Rules and Laws can lead to understanding and control of the inanimate matter through the Chemistry Field and Electricity Field.

This ability of intelligent life to understand and control the Inanimate Matter in the Universe may be one of the significant purposes of the Design.

The Universe contains trillions of planets and moons all with different temperatures, pressures and conditions. All these worlds have inanimate matter where the elements exist in solid, liquid and gas states dependent on the Fields, namely temperature fields and pressure fields that exist there. Obviously, the Design allows for lifeforms to come into existence but it must be an extremely difficult if not a miraculous process that allows for development of further Fields, Rules and Laws to keep life progressing forward to intelligence, consciousness and control of inanimate matter.

Each planet or moon has its own unique set of Fields these include –

- Position in space relative to the Star
- Orbit around the Star
- Raw materials
- Atmosphere
- Internal molten core
- Magnetic field
- Gravitational field
- Rotational period
- Axis Tilt
- Stability

All the Fields set up a unique set of laws and rules for a range of biomes for the developing life to adapt to. As we exist on the planet Earth we have proof and evidence of the following conditions and biomes for life to come into existence.

Earth's crust consists of:

A) Oceanic crust consisting of the five oceans the Pacific, Atlantic, Indian, Arctic and Southern.
B) Continental crust consisting of the seven continents Africa, Asia, Oceania (Australia and New Zealand), Europe, Antarctica, North America and South America. The Earth's crust receives different temperature and rainfall. There is also a different distribution of water and climate determines the type of vegetation found in a region. In turn the vegetation determines the type of animals found in the region. A biome is a region of the Earth characterized by climate and containing distinctive plant and animal life. A biome is made up of smaller distinct regions called habitats. A habitat is defined as an area in which

an organism lives. In any habitat, a number of species depend on one another. Plants provide food and shelter for animals. In return, animals may help to pollinate the plants. The plants and animals form a community where every habitat provides the right conditions for the plants and animals that live in it. Types of biomes currently determined by scientists include:

1) Desserts are very dry regions where only a few plants grow. They have temperature extremes either hot or cold.
2) Grasslands are the most common in temperate regions. In tropical regions with a long dry season, the typical grassland is savanna, grassland with scattered clumps of trees.
3) Oceans form by far the largest biome in terms of extent. The species that live in a given ocean habitat are determined by the depth, sunlight penetration, temperature, water conditions, and nutrient availability in that location.
4) Scrublands are areas where bushy forms of vegetation dominate. Summers are hot and dry, and fires are frequent.
5) Taigas also called boreal forests, are regions of subarctic coniferous forests. Winters are cold and long.
6) Temperate forests are found between the tropical and polar regions. The climate is mild with moderate rainfall. Temperate forests may be coniferous or deciduous.
7) Tropical rain forests grow where the weather is hot and humid all year. They form the richest biome in terms of its variety of plant and animal species.
8) Tundras are cold, dry regions where the subsoil is permanently frozen.

Earth's atmosphere is very important because in contains the gases that support plant and animal life. It also holds the water vapor that goes to form the clouds and rain, so important for the water cycle and of course the processes of weathering that dissolves the elements in the rock and carries them to the oceans as nutrients for life. The air pressure is significant. Steady high pressure indicates clear weather, while a fall in air pressure warns that stormy weather is on its way. In addition, wind, fog, mist, snow, rain, clouds, air pressure, thunderstorms, lightning, tornadoes and hurricanes are all weather phenomena generated in the Earth's atmosphere some particularly designed for bringing about stable and calm equilibrium conditions on Earth's surface.

Life must start out as simple lifeforms, and the rules and laws that develop as a result of the Fields determine or control the lifeforms development.

To survive the lifeform must follow the following rules –

- Convert the energy of the star into its internal energy.
- Find raw materials like food and water.
- Process the raw materials within its body to extract the energy.
- Transport the energy around the body via a liquid.
- Control its internal temperature.
- Excrete waste products.
- Reproduce.

Now, the planets position around the star determines the star's energy or radiation available for its lifeforms.

The planets orbit around the star determines the relative increase or decrease in the radiation available to its lifeforms.

The planets raw materials including liquids and solids determines how easy it will be for the lifeform to convert materials available into food and then into internal energy.

The planet's atmosphere determines how the star's energy may be trapped or released. It provides a source of fuel as oxygen does in the air. It also, sets rules regarding flying and motion on the surface of the planet.

The planet's internal molten core determines the planet's internal heat release. If the planet is cooling rapidly, the internal heat release could stop runaway external cooling. The planet's internal heat release is capable of supporting life when external heat is not available.

The Planets Magnetic Field driven by a rotating molten core provides protection from harmful radiation from the star.

The planets gravitational field sets the size limits for the lifeform and the lifeforms ability to move. Planets with relatively strong gravitational fields would have small creatures while planets with relatively weak gravitational fields would have bigger creatures.

The rotational period of the planet determines it's day and night cycle. Also, this Field determines the planets daily heating and cooling periods.

The planets axis tilt together with its period of revolution about the star determines the planets yearly seasonal changes.

Stability is a very important factor. Firstly, the planet needs to receive stable energy from the star. The design allows the stars to shine for billions of years in a stable manner with nuclear fusion. Secondly, the planets orbit and conditions need to be such that it has relatively natural conditions that prevent its temperature and pressure from varying drastically as to make living conditions very extreme for long periods of time.

All the above conditions together with the lifeforms requirements to survive and reproduce described above create the Fields, laws and rules to make lifeforms have:

- Sensors.
- A system for converting raw materials and starlight into energy.
- A system for extracting energy from the air.
- A system for excreting waste.
- A brain for controlling everything within the lifeform.

If it is outstanding for the Creator of the Fields, laws and rules to change the dust in Space into Galaxies, Dark Matter, Stars, Planets and Moons where atoms can exist and form compounds which can have solid, liquid and gaseous states. Then, it is more outstanding for Creator to create life that survives and reproduces given starting chemistry of basic atoms forming molecules and compounds.

On the Planet Earth we see the above, where life can survive and reproduce in near perfect conditions of temperature and pressure to support the multiple processes required for life to exist. It is truly magnificent how single cell creatures could become the lifeforms we see today namely bacteria, viruses, trees, plants, fish, sea creatures, insects, birds, animals and humans. Each of the above categories have millions of different species, types and sub-categories almost as varied as the galaxies seen through powerful telescopes.

Is there a Design behind the lifeforms that make them exist or how can Rules, Laws and Fields cause life to exist? Lifeforms need to convert sunlight or a planets raw material into its internal energy to survive for its lifetime, and a lifeform needs to reproduce copies of itself before it dies, so that it can continue to exist as part of the next generation. Life on Earth is designed to do the above in not only a variety of ways but in the most extraordinary manner.

The Animate Field has rules and laws which generate a range of sensors, some of these are as follows –

> Touch Sensors – branches and roots for trees; skin, beaks and tentacles for fish, animals and birds,

> Smell Sensors – the nose for animals.

> Sound Sensors – the ear for animals.

> Taste Sensors – the mouth for animals, birds and fish.

> Electromagnetic Field Detection Sensors – eyes for animals, birds and insects.

These Sensors generally provide Field data to a controller or brain that controls the action necessary to find the raw materials required for survival. The brain or its equivalent is designed with inbuilt drivers of feelings in the form of hunger, thirst, and love and it also has feeling indicators in the form of pain, fear, and pleasure.

So when the brain's hunger feeling is activated, it should drive the organism to find food. If the organism finds a tasteless berry and eats it, the brain experiences a sensation of disgust forcing the organism to spit out the berry and find something better like a plum. When the organism eats the plum the brain gets a sensation of pleasure and sometimes remembers the events for future use. Obviously, this is only a means of acquiring the food. The food once eaten must be digested and the energy must be used by the organism to carry out complex actions such as growing, moving, and flying around. Generally, a constantly beating heart is useful for keeping things like blood flow moving around the organism providing nutrients all over the organism constantly. Also, generally breathing oxygen from the air is useful for metabolism in animals while breathing in carbon dioxide from the air is useful for photosynthesis in trees and plants.

The finding of raw materials and transforming them into energy is not a complete design because death results in the end of things. So to survive, the Design includes reproduction. Generally, trees and plants produce seeds. Fish produce eggs. Birds produce eggs. Animals including humans produce young in a mother's womb. The extraordinary use of the sensors, the drivers, the indicators including all manner of displays and dancing is mind boggling and beyond the scope of this book. Also, the manner in which the data is stored in seeds and cells to reproduce the new organism with variability similar to the formation of a new planet or moon, is equally mind boggling and beyond the scope of this book.

The Fields, Rules and Laws allow reproduction and variability. Reproduction is required to continue the species in time, and obviously variability is required because the System is open and conditions can change. Variability allows the species to adapt to changes in the Fields, Rules and Laws. Variability is necessary for long term survival such as climate change or habitat change.

The following animate lifeforms exist in the biomes of the Earth:

1) Fungi, mushrooms, toadstools, yeasts and slime molds.
2) Lichens.
3) Plants with roots, stems and leaves.

4.1) Non Flowering Plants – Ferns, mosses, liverworts with no flowers or seeds dispersing spores on air currents.

4.2) Flowering plants – Produce seeds in the ovaries of their flowers, after fertilization, fruit is formed to protect the seeds.

5) Trees – have elaborate root systems to find underground water. Huge tree trunks to take water up to leaves. The leaves are supported on tree branches which have the ability to detect sunlight and grow strong enough to support the leaves in severe storms and harsh weather conditions. Like plants, the trees make their own food via a process known as photosynthesis. Water from the roots and carbon dioxide gas from the air combine to make glucose (a sugar) and oxygen gas. The trees use glucose as a fuel to make energy in a process called respiration. Glucose molecules are joined together into long chains. One chain called cellulose, is used for growth and developing strength. Another chain, called starch is used as a reserve food store. Plants also make amino acids for proteins, enzymes and hormones. Trees protect their seeds by encasing them in fruit. Trees do not choose where their seeds spread. Animals and insects disperse the seeds by eating the fruit or pollinating the flowers as the case may be.

6) Marine Invertebrates – these are the first multicellular animals with no backbones and they live in the seas and oceans. The oceans are still full of invertebrate animals. This group includes corals, sponges, jellyfish, starfish and anemones.

7) Mollusks-these creatures have soft bodies covered by a mantle. Mollusks live in the water and on the land. This group includes clams, snails, slugs, octopuses and squid.

8) Worms – these creatures are legless invertebrates that live in the soil or in the water.

9) Crustaceans – this group is known as arthropods and contains 38,000 species, many of which live in the oceans. This group includes woodlouse, crabs, shrimps, prawns, and lobsters.

10) Spiders, Centipedes and Scorpions – These animals are arthropods, like crabs and insects. Spiders and scorpions are known as arachnids and centipedes and millipedes are myriapods.

11) Insects – This group includes bees, flys, mosquitoes, wasps, dragonflies, grasshoppers and butterflies. They can live anywhere on the Earth and eat any food, but can only grow a few inches long. These creatures lay eggs, and are capable of amazing transformation. The butterflies and moths go through a four stage life cycle namely egg, larva, chrysalis and adult. The caterpillar hatches from the egg and eats plants. When fully developed it spins a silk cocoon known as a chrysalis from which a fully formed butterfly emerges to fly away, find a partner and repeat the cycle by laying eggs. Grasshoppers and cockroaches go through a three stage cycle, namely egg, nymph and adult. The nymph is a smaller version of the adult, though it cannot fly.

12) Fish – this group are the first animals to have backbones (vertebrates) and skeletons of bone. Fish have brains, eyes, blood, spinal cord and a heart. They take water in through the mouth and expel water out via the gills. The gas exchange is complementary to the trees. Fish take in oxygen and give out carbon dioxide. Fish are cold blooded.

13) Amphibians – are the smallest class of vertebrates. They were the first animals to colonize the land. However, they have to return to the water to breed. Most amphibians begin their lives in the water and breathe with gills. As they grow they develop lungs and legs and are able to move on dry land. Amphibians are cold blooded. These creatures include newts, salamanders, frogs, toads and caecilians.

Generally, females produce masses of eggs or spawn, which are usually fertilized externally. After about ten days the eggs hatch into tadpoles with gills. The tadpoles live and feed in the water. Eventually, the tadpole grows back legs, then front legs, and its tail shrinks. Finally, after the adult frog has developed lungs, it clambers out to find food on dry land.

14) Reptiles – are cold-blooded animals that usually prefer to live in warm climates. They are characterized by their dry, scaly skin. Most reptiles lay leathery-shelled eggs on land. The shells prevent the embryos from drying out. Reptiles are believed to be the dominant lifeform for 150 million years, the best known of these ancient animals are the dinosaurs. Today there are four main groups of reptiles:

14.1) alligators and crocodiles (about 25 species)

14.2) tortoises and turtles (about 250 species) – characterized by their hard shells which protect against attack.

14.3) snakes (about 2,700 species)

14.4) lizards (about 3,700 species)

15) Birds – are the largest group of warm-blooded vertebrates. All birds have feathers, beaks and two front limbs that have been modified into wings.

16) Mammals – are warm blooded animals that feed their young from milk-producing glands. Mammals are extremely diverse and can be found in almost every habitat on the Earth. Mammals belong to the class Mammalia and consists of three main groups:

16.1) Monotremes – these creatures lay eggs and include species like the platypus and echidna.

16.2) Marsupials like the kangaroo give birth to partially developed young. As a result the offspring stay in the mother's pouch and feed on her milk until fully developed.

16.3) Placental mammals – this group gestates (develops young) inside their bodies, providing nutrients through the placenta in the uterus. These creatures include flying mammals like bats, sea mammals like seals, dolphins and whales, large herbivores or plant eaters like elephants and giraffes, and powerful carnivores like dogs, cats and bears. The primates which include apes, monkeys and human beings also belong to this group.

Planets and moons revolving around a star are places where the temperatures and pressures create rules and laws that result in a crust or surface layer forming rocks and minerals with elements and compounds that can exist in solid, liquid and gaseous states. The planets and moons can become active or dynamic where the surface can change. Earth has an extremely active surface. Volcanoes constantly change the surface adding new material. Water exists as solid ice, liquid water and gaseous water vapour. This changing of state together with the rapid movement of liquid water transforms the surface crust and creates all the different forms of rocks and minerals. The large oceans on Earth's surface become places where all sorts of elements and compounds are brought together so that the rules and laws of life could form and become more complex. Over long periods of time the rules and laws formed the biomes. Again over long period of time the animals and plants began to coexist in the different biomes. As a general rule, the soil and animal life of a region are closely linked to its vegetation. Biomes are usually named after the main vegetation, such as grassland or forest. As a general rule, the vegetation is closely linked to climate, so biomes correspond to climate zones. Many plants and animals have features that make them especially suited to a particular biome. An ecosystem is a community of living things interacting with each other and their surroundings. As a general rule, when vegetation colonises an area, the first plants to grow are small and simple such as mosses and lichens. These plants stabilize the soil so that bigger more complex plants can grow. This is known as vegetation succession.

Rainforest ecosystems cover only 8% of the world's land, yet they include 40% of all the world's plant and animal species. Generally, green plants are autotrophs or producers as they make their own food. Generally, animals are heterotrophs or consumers. Primary consumers are herbivores or plant eaters. Secondary consumers are either carnivores or meat eaters and omnivores or plant and meat eaters. Generally, the decomposers are bacteria and fungi that feed on the remains of dead plants and animals. They prepare the soil for plants by decomposing the dead plant material into nutrients. Every living thing has its own unique role in the living community. The rules for lifeforms are complex though they can be simplified for the purposes of this book as follows:

The decomposers prepare the soil for the plant roots. Water in the form of rain is supplied to the soil and to flowing streams to keep the soils moist so that the nutrients can be sucked up by the plants to the leaves. Sunlight is provided every day as an energy source to allow the plants to produce energy via photosynthesis and grow. The primary consumers or herbivores have abundant plant food so that they can grow and multiply. This provides abundant food for the secondary consumers or carnivores or omnivores, so that they can grow and multiply. Finally, every living thing is made up of tiny building blocks called cells. Most cells are microscopic. They can convert food into energy, make useful chemicals, and reproduce themselves. Cells can reproduce in one of two ways. Mitosis is a process that creates perfect copies. It is used during body growth and to replace damaged or dead cells. Meiosis is a process that combines half the chromosomes from a female egg and half the chromosomes from a male sperm. It is used to produce a new offspring.

Each cell has its own set of instructions that tells it how to create the chemicals vital to life and how to put them together. These instructions called genes are found inside long, twisted molecules called deoxyribonucleic acid or DNA. DNA is coiled into structures called chromosomes. The DNA molecule forms a long, winding spiral ladder shape that is called a double helix. The ladder's rungs are made of pairs of chemicals called bases. The order of the base pairs is in the form of a code that spells out how to build proteins and other chemicals necessary for the lifeform. Sections of DNA can unzip down the middle. When this happens, each half can be used to build a new identical molecule. The simplest lifeforms contain just one cell and the most complex lifeforms contain millions of cells following the rules of their DNA code. The whole process of both mitosis and meiosis is obviously a very important part of the design because it allows life to continue and new life to be created with variability because in meiosis it combines the genetic material from two separate beings. For people in the Religious Field this part of the design strengthens the belief in a Creator because every coding process requires a code maker. The meiosis process involves two beings and tends to bring two separate parties, groups, tribes and sometimes clans together in a partnership that encourages communication and responsibility. For human beings, this part of the design has created the family unit, because in most cases families live together and encourage and influence each other as they join the Fields of Knowledge.

Having a brain that can store information is extremely important and this appears to be what human beings possess. For human beings information storage has gone from writing things down in stone, to writing on paper, to writing in books. Books are stored in libraries. Data must not only be stored it must be accessible to other beings. The more accurate the storage of data and the more accessible the data the better. Of all the animate lifeforms mentioned only human being can store data and make it easily accessible. Hence, only human beings could follow the rules and laws to create the Fields of Knowledge.

# 8
# HUMANITY

Knowledge was everywhere, but the Fields, Rules and Laws had to come into existence before that knowledge could be understood. In the last chapter Animate Matter Fields had developed on the Earth and animal life possessed sensors and a brain to find the food and water necessary for survival and reproduction. Thankfully, the rules and laws allowed humanity and the human mind to develop of Fields of Knowledge which allowed humanity to discover many of the rules and laws that are being followed by Mass and Matter in Space. These discoveries are not an accident. As far as we know, it is only humanity that believes and prays to a Creator. This belief had a profound effect in domesticating animals and growing crops that were seen as the Creator's creations. The domestication of animals and the growing of crops meant that human beings could live in villages where food and water was readily available. Living in villages provides a more settled lifestyle and created a need for language, song, dance and art within a village culture. The need to store data like food and wine recipes, and the location of food and water resources resulted in a start for the Field of Knowledge. As far as we know, no other species followed the same rules and laws associated with a belief in a Creator. Many of the early human villages have some sort of a church or religious institution. No such institution can be found among all the other species of animals and plants. In fact, the Religious Field dominated human life creating many of the rules and laws generally associated with morality and good charitable works which helped humanity and encouraged the development of the Fields of Knowledge. Fanatical religious wars are considered extreme events based on anger, usually when prayers to the Creator appear to be unanswered or when different religious institutions challenge each other. These extreme events are not part of the scope of this book. It should be noted that most religious events resulted in less needless preying on other lifeforms and more praying to the Creator, who obviously cared about the lifeforms created. The book suggests that as human beings, we should all stop needless preying on other lifeforms and pray to the Creator instead, to help us to develop the machines through the Fields of Knowledge, that will help us lead more comfortable and responsible lives that are in harmony with the biomes on the Earth. This should help us to take our consciousness and the Fields of Knowledge to other stars and galaxies. Eventually, we would have a Universe with mind and matter everywhere.

Animals developed ears that provided sound inputs to their brains. The sound inputs generally provided a warning to the brain that a dangerous wild animal was near. Reflected sound could also be used to detect the presence of objects and possibly food. The Human Mind was able to use sound to create language. This meant that two or more humans could talk to each other. This human ability to communicate meant that humans could form communities and plan future events. This would lead humans to develop villages with language, art, song and dance. In settled village life planning was necessary as food had to be provided for a number of humans and their animals. Planning would involve developing agriculture and the domestication of animals, fish and birds. Also, water had to be stored and used carefully to ensure that it did not run out during periods of drought. The Field of Language with its rules and laws is very important as a means of communication to implement the planning for the above.

Animals, fish, birds and insects developed eyes that provided visual inputs to their brains. The visual inputs generally provided a warning to the brain that a dangerous wild animal was approaching or a friendly animal was approaching. Visual information was excellent for seeing food and water. The Human Mind was able to use vision to create writing and drawings. Both these Fields are essential for planning and keeping a long term record of events. This made trade possible, villages could just do one thing like grow a certain crop and they could trade with another village that raised cattle and the whole transaction could be recorded and witnessed, avoiding violence and aggression. Also, recording data was extremely important for passing information to the next and future generations. The Field of Writing with its rules and laws is very important as it would have led to books and libraries where data and information could be stored for future generations. Before writing, data was lost or corrupted within a couple of generations, so accurate recorded events were not passed on to future generations.

Animals, birds, insects and fish developed mouths with taste sensors that provided sweet and sour sensations to the brain. The taste inputs generally provided food and water quality information to the brain. The Human Mind is capable preparing food and wine that is served as a meal. This made eating and drinking a very safe and healthy exercise. Nutrition is very important for human development. Things like washing the mouth not only made humans healthier and stronger it improved the quality of life. The Field of Cooking and Food Preparation is important and Language and Writing Fields are again important for storing this data for future generations.

Most animals have four legs for running from wild animals or for chasing prey. Only monkeys are two legged, however they show no inclination to use tools. The touch sensors of their hands are useful for climbing trees to evade predators and to find food. Humans have two hands and two feet. Hands are useful for growing crops, raising animals, caring for young humans, using tools and writing. The rules seem to encourage human brains to use their hands to carry out these essential activities.

The Inanimate Field is full of elements and minerals usually locked up in rocks. No other animal has been able to understand the significance of these elements and minerals. Human beings are able to see the ornamental significance of elements like gold and silver, and gemstones like diamonds, emeralds, rubies and jade. Also, human beings are able to extract and process the elements from rocks and minerals to produce products like houses, bridges, roads, airports, railway lines and machines essential in the modern world.

The Human Movement Field led to a Sporting Field which became an important pastime and recreation activity until it became an important money making activity in the modern world with such sports as golf, tennis, cricket, soccer, football, basketball and baseball all creating their own Fields and stars.

The Fields of Knowledge continued to increase. The accuracy of information in the Fields of Knowledge improved dramatically with improvements in storage of data and easy access to data through libraries and the availability of books and improvements in printing.

For human societies, the original drivers of hunger, thirst and love were driven into the background and replaced with getting an education and learning skills to do a job and earn a wage. The human life cycle became birth, primary education, secondary education, tertiary education (if applicable), working career, and retirement. Working career implied getting a wage to buy food and water to satisfy the primary drivers. Working career also meant finding a partner for reproduction. The initial drivers remained, however they were satisfied via education and working rather than hunting and searching for limited food and water resources. The available food and water resources were made available to more people in a civilised manner which resulted in a more organized way of life. This process resulted in the rapid development of all Fields of Knowledge and also people could select the Fields of Knowledge they liked. Interestingly, human stars began to come into existence very rapidly in all the Fields of Knowledge greatly improving the Fields and increasing the available knowledge in the Fields for later generations of humans. The process once started was continuous and resulted in rapidly evolving Fields of Knowledge. Some humans were able to form new Fields in a system that was continually changing in an 'open' sort of way. The system depended on where the human stars were forming in the various Fields of Knowledge.

The Transportation Field is very significant for humans because of its importance for the primary drivers of hunger, thirst and love. All animals have to walk or crawl to find food, water and a partner. Humans have a Transportation Field which speeds up the process. The Transportation Field was developed by inventions through education and work. Inventions like the bicycle, motor cycle, car, trains, aero plane and ships were all rapidly developed to enable people and goods to be transported. When cars were developed in the Transportation Field a whole new series of rules, laws and regulations were required to implement safety regulations and to make cars available to the community. The following had to be carried out or allowed for:

1) There had to be fuel to drive the car.
2) There had to be roads to drive the car on.
3) There had to be traffic rules to operate the car in a safe manner.
4) There had to be a police force to enforce the rules.
5) There had to be rules to allow licensed people to drive the car.
6) There had to be a mechanic to fix the cars.
7) There had to be manufacturers to manufacture the car to certain standards.
8) There had to be standards.

Generally, the Field automatically creates the rules and laws necessary to maintain its existence. If for instance item 4 above was missing. Namely, a police force did not exist to enforce the rules, then accidents would increase. People would determine that accidents were occurring because traffic rules were being disregarded and no one cared. The system would become disordered and order would only be restored by having an Authority available to implement the rules.

Hence, the rules and laws impose order in the Field. With time more and more issues occur which tend to make the system disordered. Again, further rules and laws change the disorder into order. It is interesting that the disorder is noticed by the human brain or mind and again it is the human brain or mind that add rules and laws to establish order.

With rules and laws in place, the Transportation Field still needs work and money before human beings and goods can be transported anywhere in the world. The same sort of thing with minor variations occur in all the Fields of Knowledge. Namely, one has the Field and the rules and laws that impose order. When disorder occurs in the Field, the disorder is examined and if necessary new rules and laws are implemented to correct the disorder and bring about a better more ordered system. Sometimes factors within the Field can change causing the disorder, namely a shortage of petrol. In this case alternative sources of fuel have to be found. This results in new inventions like electric cars or solar powered cars. These result in new rules and laws and perhaps a changing or evolving Field where cars are driven not by petrol but by electricity, solar energy and batteries.

All Fields of Knowledge appear to have one thing in common they need a Mind or Brain to create the rules and laws that turn the chaos into order.

Order seems to be an essential aspect of the Transportation Field.

Order is important for all Fields of Knowledge, however some Fields appear to be focused on winning and losing. The winners generally become famous stars. This is essential to the Sporting Field. There are various Sporting Fields namely swimming, running, walking, soccer, football, cricket, hockey, basketball and baseball. Each Field has a specific set of rules and regulations. Human beings play a particular sport in accordance with the rules and regulations. Some people are repeatedly successful. Some people are repeatedly unsuccessful. Those that are repeatedly successful generally become famous stars. This type of Field is an Entertainment Field, where the so called star has a following and supporters that provide the status of the star. The greater the support and following the bigger the star.

The Fields with their rules and laws impose order and are very significant because they have resulted in the development of a more ordered World. That is –

- A more ordered World of towns and cities located in states and countries.
- A World that is governed by governments that encourage rules and laws that control all the numerous Fields and resources within the country.

In an Open System, the Inanimate Matter or molten material cools to form various types of rock dependent on the heat and pressure. This is not a problem. We have also seen how ancient dead plant life can be transformed into a rock called coal. However, in an Open System there is a major problem for animal lifeforms. For Animate Matter to get food and water and to reproduce, results in a depletion of resources. This makes it harder to find the food and water necessary to live and could lead to violence and aggression. All this is enhanced by rapid reproduction or a rapid growth of the species and difficult times of famine and drought. Even this, has been allowed for in the design with a sense of morals seen in some animals and seen in detail in the Human Species.

Following the rules, animals and planets lived in the natural biomes found on the Earth. The biomes had rules where mosses and fungi prepared the soils for trees and plants. Plants and trees made their own food. Animals were either herbivores or omnivores. In some cases animals preyed on other animals creating turbulence and disturbances within the biomes. Human beings prayed to the Creator. This created an atmosphere of caring for the Creator's creation. Destroying plants on a large scale for food, was converted into caring for plants and growing crops. Hunting and killing animals on a large scale was converted into caring and domesticating animals. Praying to a Creator gave rise to a strong Religious Field that led to the formation of villages with a church and domesticated plants and animals. Villages traded with other villages leading to the development of the Language Field.

Human beings found it easy to express themselves by doing drawings or by dancing or by singing. This led to the development of the Art Field, Dancing Field, and the Music Field. As village communities grew things needed to be recorded. This lead to the development of the Writing Field.

It was like a strong sense of the Creator had introduced morals that had transformed humans from wild hunters killing animals and destroying plants into peaceful settled groups growing crops, and raising animals. Then, these ways combined with writing, language, and music led to civilization and cultural development with great improvements in the Fields of Knowledge. Small civilized groups could trade goods produced and this helped to improve lifestyles and add variety to the Fields of Knowledge.

Since morals were so important in the development of civilised human life, perhaps raising humans above the level of animals and other creatures, I thought I might see if I could find the supernatural through moral codes and laws. The source of moral behavior seemed to be contained in ancient religious books like the Bible, the Koran and many other religious books.

On investigation, I found that the Bible contained things like –

The Ten Commandments.

Moral teachings including the teachings of Jesus.

The story of Adam and Eve and the forbidden fruit indicates that rules and laws may be broken but the breaker must pay for his actions.

The story of Cain and Able indicates evil and good in the World.

Unfortunately none of these religious books contained scientific proofs, they were all based on faith. However, they set the framework for the development of the Fields of Knowledge, by allowing humans to develop a more varied lifestyle through civilization and trade with other human groups.

With writing and paper, human beings could accurately record information and events. Paper had limitations and it was laborious to rewrite documents and data when the paper deteriorated. Lately even this issue has been resolved with computers and electronic recording.

Thus, in the modern world data and knowledge can be easily stored and added to.

In fact, one the purposes of a human life may be seen as living and adding to the existing Fields of Knowledge, then the new Fields of Knowledge are opened to the next generation for adding to and developing and improving the Standards. This is what is happening to the Human Mind and the Fields of Knowledge. This is similar in a manner of speaking to what is happening to Matter and the Field of galaxies. In the Field of the galaxy the stars are born, they live and die ejecting the elements into space and becoming the Dark Matter that can shape the galaxy and allow for the formation of new stars and new galaxies.

# 9
# FIELDS OF KNOWLEDGE

As far as we know, the Human Mind is the only thing that is aware of and is capable of creating and improving the Fields of Knowledge. Certainly, one of the purposes of Humanity seems to be to develop, create and add to the Fields of Knowledge.

Through the Fields of Knowledge we have seen how the tiny atoms of the Periodic Table of elements were created in the stars and distributed to the numerous planets and moons that revolve around the billions of stars. The process of nuclear fission has shown the enormous energy stored in the atoms, where energy is approximately equal to mass times the speed of light squared. The atom forms the basic unit of the solid objects we see around us on earth. In a similar manner, we have seen how the Star System with planets forms the basic unit of the galaxies we see in the telescopes. The Fields of Knowledge cover an unimaginable range of Fields. Fortunately, though our senses are unable to detect both atoms and galaxies, the tools available through the Fields of Knowledge enables to see, hear, touch and sense almost everything in the Universe. This is amazing, but it is also something that requires an amazing amount of work to sustain and maintain. Fortunately, like the generations of stars continue to support the billions of galaxies, there are generations of humans that will need work to maintain the Fields of Knowledge for generations to come.

We have seen the rules and laws that form the two purely Human Fields of Transportation and Sport. The Transportation Field shows that it is not only possible to move people and goods at phenomenal speeds on the Earth, but it is possible to move beings and goods out into Space, if necessary. The Field of Sport is a competitive Field and is mainly for entertainment and athletic improvement.

The Fields of Knowledge belong to the Human Mind and are composed or developed by living and dead human stars that are as vast as the living and dead stars and associated planets, moons, debris and dust, that compose or make up the Matter in the Fields of the galaxies. Humans use the senses of sight, hearing, smell, and touch not to search for food and water but in the vast Education Field. The Education Field contains the teachers and professors in numerous schools, colleges, universities, and education centers that teach information about the Fields of Knowledge at primary, secondary, tertiary and higher levels.

After education human beings are exposed to the variety of Fields that exist in the modern World. Some of these Fields are as follows:

1) Transport Field includes cars, bicycles, ships, trails, aero planes, motor cycles, trams and rockets. This is an enormous Field with numerous rules and laws some of which have been described above.
2) Sports Field – This Field involves exercising all parts of human body and developing the brain to carry out specialized movements and co-ordination. Human beings can compete in individual sporting events or team sporting events. The number of sports is unbelievably large. Sporting events usually involve winning and losing. Human beings are often remembered an honoured for outstanding displays like breaking a record in swimming or athletic event.

## Fields of Knowledge

3) Water Industry Field – This Field is essential for human health. Clean, healthy water is essential for life as we know it. This Field involves making water available for human and animal consumption. It includes storing water, treating water, cleaning water and distributing water.

4) Air Industry Field – This Field is essential for human health. Clean, healthy air is essential for life as we know it. This Field involves making clean and healthy air available for human and animal consumption. It includes filtering the air and stopping pollution of the air as occurs in factories, underground tunnels and through the excessive use of transport vehicles.

5) Energy Production Field – This Field is essential for modern city living. This Field involves finding natural resources like coal, oil, gas, uranium, water, solar, and wind and transforming them into energy. Then distributing this energy in a safe and reliable manner such that it can be used by individual human beings or groups of human beings and animals.

6) Food Production Field – This Field is essential for civilized living in city and country areas. This Field is vast as humans and animals need large quantities of food to survive and work in the Fields themselves. This Field involves crop production, animal caring and production of animal products like cheese and milk. This Field also involves making plant and animal produce available to individual human beings. A big task when city populations are in the tens of millions.

7) The Clothes and Grooming Field – In the civilized world all human beings wear clothes, shoes, have haircuts and women have additional makeup sessions. This includes the manufacture and selling of clothing and shoes plus all the associated grooming industries including facial and nail manicures.

8) Manufacturing Goods Field – This Field is essential to the human lifestyle. Most humans not only like fast transport but they like goods like refrigerators, cooking appliances, washing machines, dish washers, heaters, air conditioners, shavers, and

FIELDS

thousands of associated appliances. This field produces and delivers these important items.

9) Building and Construction Field. In chapter twelve the design of a house was considered. It is essential for human beings to live in separate houses or units, so that, they do not fight over the complexities of life in the many Fields. The Building and Construction Field includes obtaining building materials from raw materials found on the Earth. It is important to note that many of these substances like everything else on Earth was created by the rules and laws during the creation of the Solar System. Since there are millions of buildings in millions of countries this appears to be a significant amount of raw materials and a creation process for which we should all be grateful.

10) Mining Field – This Field absolutely destroys the Earth that early humans once knew. To sustain large populations it is essential to explore and mine the planet to find the necessary materials to produce the energy and raw materials required to live a normal civilized life.

11) Religious Field – It is important to remember the Designer or Creator as some prefer, who made everything possible. It is also important to help fellow human beings and animals because when resources are used at the rate required to exist in the modern world, hardships and disasters are bound to occur. This Field is essential because of the complexities in the Fields themselves.

12) Medical Field – The senses seem to have a 'use by date', obviously they cannot last forever. The general time of use is 80 years from the date of birth. Hence, the medical field is essential for the repair and maintenance of the senses, the body parts and the brain. One cannot do without this Field it is essential for life.

13) The Communication Field – This Field is very significant because the development of listening and speaking in human beings made significant changes to the use of the senses, as

mentioned in the earlier chapters. The Communication Field includes the development of thousands of languages and led to development of thousands of dances accompanied by music and songs. It is easy to see how this Field helped in social development and contributed to such things as love and reproduction. The importance of this was mentioned earlier. This Field is however more important, for reliable data transfer and storage. Modern communication systems do all these in a very safe manner complete with automatic error checking and automatic error correction. This Field includes automatic data transfer at a distance using television networks, radio networks and telephone networks. The above is accompanied by use of computers, the internet and the World Wide Web. Even newspapers, which were a very important means of passing current affair data among human beings, have become secondary. Information on the Internet is available complete with video data of the events.

14) The Spaceship Earth Field – The whole Earth is moving with the Solar System around the Galactic Centre at an approximate speed of 800,000 km/hour and will take approximately 220 million years to go around the Galactic Centre. The Earth could be considered as an ark travelling through Space and Time. Therefore, this Field encourages caring for the other plants, animals and sea creatures, because if lost they may never return. The caring help is possible through the Scientific Field.

15) The Jewelry Field – In an earlier chapter we saw how the rocks in the Earth's crust are subjected to varying temperatures and pressures resulting in the creation of gemstones. This is a huge Field involving the mining and working of gemstones like diamonds, opals, emeralds, rubies and sapphires for display and sale with precious metals like, platinum, gold and silver.

16) The Food Preparation Field – Human beings do not eat things raw like all the other creatures that inhabit the planet. The senses of smell and taste create a human appetite that is usually

satisfied by well flavored, exotic, tasty and well served meals usually served with entretrays, main meals and desserts. This is something no other creature on the planet wants or gets. Yet it is something human beings insist upon and it is essential for a civilized lifestyle.

17) Legal Field – The laws and rules in all the above Fields make this Field essential. This Field enforces the rules and laws by exercising penalties, including fines and jail terms for offenders. It is linked to the police and army for more serious offences and off course war when other countries wish to expand or exert dangerous influences over neighbouring countries on the Earth.

18) The Political Field – This Field is associated with the Government within a country and is generally run through elections over a three or four year period. The Government makes the rules and laws to govern the country and its people. This Field is closely linked with the Legal Field described above.

19) The Business and Economics Field – Nothing is complete without this Field because human work, goods and services must be converted to a monetary sum. Work, goods and services must all be paid for and taxed in a proper way otherwise total disorder and chaos could easily occur. Goods and services must be priced such that they can be bought with the wages people earn. Trade with other countries is a more advanced form of initial trading that occurred in early societies resulting in specialization and an easier lifestyle. This Field includes setting up of basic wages within individual countries and exchange rates among different countries. Computers are essential for monitoring and recording a vast number of transactions that occur constantly. This Field can be mind boggling as millions of prices rise and fall on a daily basis. Many of the transactions are repetitive or similar, however investors make this Field very uncertain and difficult to predict.

Fields of Knowledge

The Scientific Field is part of the educational field that is important for storing knowledge and data. It became significant after the 15$^{th}$ Century AD and was looked at in parallel with the Religious Field because of the 'open' nature of the Design. This Field falls roughly into six groups:

a) Mathematics – the study and relationships between numbers – This includes computer science that investigates computer programming and how computers run their programs including how inputs and outputs both analog and digital are processed.

b) Social Sciences – using scientific methods to explain how humans behave on their own, in groups, or in society. This field includes:

    B1) Psychology – The study of the human mind or psyche and why people behave the way they do, often to help people cope with mental stress and illness.

    B2) Archaeology – Investigating traces of human life in the ground, including bones, tools, and ruins, to figure out how people lived in the past.

    B3) Human Geography – Researching how humans use the landscape, including the patterns of their settlement, transport and activities.

    B4) Economics – Examining how people, companies, and governments decide which goods and services to produce and consume.

c) Biological Sciences – Studying things that are alive, or were once alive. These sciences are also known as the Life Sciences. This Field includes:

    C1) Epidemiology – tracking the spread of diseases, including fast-spreading epidemics, from person to person, through a community, and around the world.

    C2) Pathology – researching diseases, how they attack the body, and how they can be stopped.

- C3) Medicine – Finding cures and treatments for illnesses and injuries, generally using chemical medicines and physical treatments such as surgery.
- C4) Anatomy – Figuring out the structure of the body and how it works.
- C5) Botany or Plant Science – Studying plants, including their variety, structure, chemical processes and life in the wilderness.
- C6) Zoology – Studying animals, including how they survive and reproduce in the wild.
- C7) Microbiology – Examining the smallest living things on a cellular basis using microscopes to see the details.

D) Combining the Social and Biological Sciences – This Field combines the Social and Biological sciences:
- D1) Anthropology – Investigating the variety in the human species, including how humans and their different cultures have changed.

E) Earth Sciences – Studying the non-living parts of the Earth, such as rocks, volcanoes, and the atmosphere. This Field includes:
- E1) Geology – Studying how Earth's rocky surface layer has formed and changed over the life of the Planet.
- E2) Meteorology – Understanding Earth's different climates and predicting changes in the weather.
- E3) Physical Geography – Studying the formation and structure of Earth's surface features.
- E4) Oceanography – Learning all aspects of oceans and seas, including tides, currents, the chemistry of the water and marine life.

F) Combining Biological and Earth Sciences; These Fields combine the Biological and Earth Sciences namely:
- F1) Ecology – Understanding how different lifeforms rely on each other and on their environment for survival.

- F2) Paleontology – Studying fossils and what they tell us about living organisms in the past, including how life began and how it has evolved.
- F3) Environmental Science – using a range of information from biology, chemistry, and geology to understand how all the different parts of the environment work together.
- G) Astronomy – This Field Studies objects in the Universe beyond Earth, such as stars, planets and galaxies.
  - G1) Cosmology – Learning about the Universe as a whole, how Galaxies, Stars, Planets and dark matter came into existence.
- H) Physical Sciences – This Field includes the study of non-living things, such as atoms, energy, and radiation. It includes:
  - H1) Optics & the Electromagnetic Spectrum – Investigating the behavior of light – its wavelengths, how it bends, reflects and scatters. Investigating the different parts of the electromagnetic spectrum gamma rays, X-rays, Radio Waves, Micro Waves, Light, Ultraviolet and Infrared Light.
  - H2) Chemistry – Studying the Period Table and how atoms connect to produce new substances in reactions and studying the properties of new compounds.
  - H3) Electronics – Examining the flow of electric currents through substances, and how to use them to run microchips, computers and other devices.
  - H4) Nuclear Physics – Looking inside the atomic nucleus. Understanding radioactive decay, nuclear fission and nuclear fusion.
  - H5) Mechanics – Studying how forces make objects move.
  - H6) Acoustics – Studying the behavior of sound waves.

H7) Engineering – Using scientific understanding to build useful machines, power systems, computer systems, water pumping systems, gas systems, and civil and architect support systems.

H8) Robotics – A field of engineering aiming to build machines that can move, sense their surroundings and think for themselves.

I) Physical and Biological Sciences – These Fields are a combination of the Physical and Biological Sciences:

I1) Cell Biology and Molecular Biology – Understanding how cells inside living organisms work both as a whole and at the level of their DNA, proteins, and other complex molecules.

I2) Forensic Science – Using a range of scientific knowledge to investigate evidence from crime scenes.

I3) Nanotechnology – Using engineering, molecular biology, and other knowledge to invent machines such as nanobots, no bigger than a human cell.

I4) Genetics – Studying how characteristics controlled by genes are passed on from one generation to another and how they vary in living things.

I5) Immunology – Examining the immune system of humans and animals. Determining how the immune system defend the body against attack from diseases.

I6) Biochemistry – Studying the chemical reactions that drive life processes.

I7) Biophysics – Using chemistry and physics to study biological systems, such as animal senses and the way animals move.

All the above Fields are populated by human beings during their working lifetimes usually after completing some primary and secondary education. There is a potential for human beings to become famous, successful stars of differing magnitudes in the Fields during a working lifetime of 5 to 40 years. The star fades but the persons influence shapes the Field and the work carried out in the Field is passed on for future generations of stars. A generation in human terms is only about thirty five (35) years. Hence, the famous stars are numerous and beyond the scope of this book. Any Field is composed of both living and dead stars. The dead stars generally shape the Field giving it validity and shape, while the living stars are generally burning their energy working in their chosen Field and enhancing it by keeping it active and alive. Some people could see stars as having different magnitude as in the Sport Field where athletic stars can achieve fame at a school level, state level, country level or Olympic level. As there are many Fields it would be impossible to list every star in every Field. As an example I have chosen five stars and one group of stars in five Fields as follows:

1) Field – Transportation

    Category – Motor Vehicles

    Person – Henry Ford

    Revolutionised the Motor Vehicle Industry making cars available to the general public. Transport by horse and cart was changed to transport by a mechanical machine. Introduced the conveyor belt production system where workers did simple routine tasks for relatively high wages.

2) Field-Sport

    Category – Soccer

    Person-Pele

    Took Brazil to three World Cup finals

3) Field – Sport

   Category-Cricket

   Person – Don Bradman

   Has the best Test Match batting average of 99.9 runs per innings. Played Test match cricket for Australia

4) Field – Communications

   Category – Painting

   Person – Vincent van Goh

   Painting many masterpieces. Not famous during his lifetime and committed suicide. Famous after his death.

5) Field – Communications

   Category – Music

   Group – ABBA

   About 20 Number 1 hit songs between 1973 to 1980. Top of the Musical Charts for a decade. Music still played today.

6) Field – Science

   Category – Astrophysics

   Person – Isaac Newton

   Described the Laws of Planet Motion and the Law of Gravity that controls objects in Space and allows water to move on Earth.

## Fields of Knowledge

Many generations ago, human beings, many of them stars in the Field of Science, Category Astrophysics considered that the Earth was in the center of the Universe and that everything, all matter revolved around the Earth and that this was the will of the Creator. We can now see that these human stars were not entirely wrong. For though the Earth is not in the center of all the matter in the Universe, it certainly is the center for the Mind and Consciousness in the Universe. On the Earth, the rules and laws allow water to flow and sunlight to provide the energy, so that inanimate matter can give rise to animate matter. When animate lifeforms believe in a Creator the animate lifeforms can develop Mind and Consciousness that lead to the development of the Fields of Knowledge which help the animate lifeforms to see beyond the horizon.

# 10

# BEYOND THE HORIZON

The Fields of Knowledge take us Beyond the Horizon because they have provided us with the technology and instruments to see beyond the horizon. The telescopes show us the beautiful macroscopic world of the stars in the galaxies. The microscopes show us the beautiful microscopic world of the elements and atoms that form the molecules, compounds, single celled lifeforms and multicellular lifeforms. We live in a complex world with complex interacting Fields all around us. It is truly amazing that we were given the brains and the mind to understand all these complexities.

As far as we know human beings are the only beings capable of understanding the Universe. No other living lifeform or dead lifeform according to the fossil record was capable leaving its home world and venturing out into the creation beyond.

Fields shows us that life is complex. The very action of having a cup of tea and a piece cake with my wife at home on a sunny summer day required a complex set of Fields to interact for the event to take place as follows:

I needed a metal kettle to boil the water and the kettle needed a cable with copper conductors to transmit electricity from a power outlet in my house to the kettle. The metallic elements of copper and tin had to be created in cores of stars. These stars had to explode to enrich the gas clouds within the Milky Way Galaxy with the heavy elements. The solar system had to form and Earth had to form in a region that captured the elements of tin and copper from the vast gas cloud. The Fields of Knowledge had to develop so that human beings could extract the metal from the rocks in the Earth's crust and process them. The Manufacturing Field had to be available to produce the kettle and cable with the copper conductors. Houses had to be provided with electrical power from a power network and a water supply from a water network. My house had to be constructed by a Building Construction Field. I had to fill water in the kettle by turning the tap. Earth's gravitational field had to follow the rules to push water from the water pipeline into the kettle. I placed the electric kettle on the electric stand and turned the electricity switch, this followed rules creating a potential difference between the copper conductors connected to the kettle. The electrons in the outer shell of the copper in the active wires were loose and followed the rules moving in the Electric Field just created. The electrons formed a current that flowed through the coil inside the kettle. Electric energy was converted to heat energy. Following the rules of thermodynamics the heat energy was transmitted to the water and after a couple of minutes some of the water in the kettle was converted to a vapour. The steam activated a pressure sensor in the kettle, which automatically disconnected the circuit. The flow of electrons was stopped. I poured the hot water into two cups, then dipped two tea bags in each cup. Almost immediately, the rules and laws sprang into action as the tea came out of the tea bags and as if seeking equilibrium it moved into the water in the cup. I poured milk into the cups and again the milky solution spread finding an equilibrium. I then cut two slices of chocolate cake and took the tea and cake to my wife. As we ate and drank, there were other rules and laws that sprang into action. These rules and laws converted the cake and tea into chemical energy within our bodies. As our hearts pumped blood automatically, this chemical energy would be transported around to the cells within our bodies, so that we could continue to be alive.

As we looked up into the sky, it was the Sun that made the day beautiful. It too was following rules and laws, squeezing atoms in its core, to convert mass into large amounts of energy and then radiating that energy into Space. It was this heat energy radiated from the Sun that kept the morning temperature on Earth at 30° Centigrade. It was the tilt of the Earth's axis and the position of the Earth in its orbit around the Sun that made it a summer day.

Finally, we needed to realize the beauty in the day. We needed a brain with sensors:

a) Eyes that were sensitive to the radiation produced by the Sun.
b) Skin that was sensitive to heat and touch.
c) Taste to enjoy the cake and tea.
d) Ears and voice to communicate.

Still the day would not have been beautiful without the Fields of Knowledge developed by human beings over the ages and in particular the development of language which made communication possible. Finally I said "It is a beautiful summer day and the tea and the cake are delicious." I heard a disturbance as the air particles vibrated at my ear drum. My brain interpreted the disturbance as two words "I agree."

Basically, there were two kinds of rules and laws that made the day beautiful, these were:

a) Manmade like the water connection to pipes in the Water Network and controlled by the tap, the power supply to the electrical outlet, the cable with copper conductors and the manufacture of the kettle.

b) Natural like the Sun shining and radiating heat, the water molecule found in liquid form on Earth, the element copper, the metal of the kettle, our brains and sensors.

The manmade rules and laws are based on the Fields of Knowledge – water is pumped from a reservoir or river, then treated and finally connected via a network of pipes to houses. Electricity is generated at a Power Station and delivered via a network of cables to houses.

The natural rules and laws are natural and have existed throughout human consciousness.

The manmade rules and laws have arisen out of the Fields of Knowledge.

The Transportation Field provides an excellent example of how the rules and laws work as follows:

We know that when we drive cars or fly planes we need to follows rules and laws like:

a) Speed rules.
b) Do not drink excess alcohol rules.
c) Obey the traffic sign rules.
d) Obey the traffic light rules.

If the rules are not followed we would get disorder and in extreme cases chaos. The rules are necessary to make order and prevent disorder.

We know how the human rules come about, but how do the natural rules and laws come about?

# 11

# DESIGN

As every designer knows, a design is a complex process. In human terms every design must have a purpose. If someone wants to live in a city suburb this means the person has a purpose. To live in a city suburb the person or family require a house. The house needs to be designed to suit the person's requirements.

The house designer or architect has to ask a detailed set of questions like the:

1) Number of bedrooms required.
2) Type of dining room and kitchen.
3) Laundry requirements.
4) Toilet requirements.
5) Floor and wall details.
6) Type of paint required.
7) Internal furnishings.
8) External requirements.
9) Electrical requirements.
10) Type of driveway and garage required.

D ESIGN

11) Type of bricks required.
12) Type of fence required.
13) Type of garden required.

Plus endless other requirements before a draft building complete with the electrical and mechanical design can be undertaken.

This is a long labourious process. After this draft design is complete, the architect has to review the draft design with the client before finalisation of the design stage. Then the construction phase is undertaken based on the requirements of the design specification.

The Universe and all it contains does not have any specifications. Observations indicate that the Universe exists and changes with time. There appear to be Fields and Rules which are followed by all the Dust and Matter. A designer seems to have set up or configured the Fields and Rules and then allowed the process to evolve and develop.

The design appears to be an open system with Black Hole Fields giving rise to Galaxy Fields containing billions of living and dead stars. The living stars pour out energy which is received by planets and moons revolving in the Star Field. The dead stars help to shape the galaxy. The open system allows for the elements formed in the cores of stars to be distributed to the planets and moons of the Star System. The planets and moons form in a molten state and cool to form a molten core with a thin crust. The crust of planets and moons contains inanimate matter. In the case of the Earth the crust of the Earth is like the skin of an apple. Inanimate matter can be subjected to all ranges of temperatures and pressures. The temperatures and pressures change the inanimate matter forming all kinds of rocks, minerals and gem stones with beautifully arranged structures and patterns which are controlled by Fields, Rules and Laws that appear to be preconfigured. Generally, hot molten rock from the planet or moons core rises up to the surface as gravity tends to pull the surface crust inward. This hot molten rock results in volcanoes and hot springs that effect other rocks because all sorts of chemical reactions can occur to transform the rocks into beautifully arranged structures that remain in the ground.

It requires a form of mind or intelligence to find, extract and admire the beauty of these structures.

Sometimes as on Earth, animate matter can come into existence in the form of cells. Cells have a very limited range of temperature and pressure at which they can survive in relative comfort. Cells are extremely delicate in most cases and need a constant supply of energy to sustain their existence. On Earth most lifeforms or living organisms require sunlight, food and water to exist. Animate matter is relatively short lived. Animate lifeforms require sensory equipment to find the sunlight, food and water to exist.

It is obviously difficult for a single organism to exist on its own, so animate lifeforms automatically form groups to enable them to survive. The more complex and diversified a group, the easier it is to survive. Again these are all subject to Fields, Rules and Laws that appear to be preconfigured or might appear spontaneously.

As a typical example one could have a Religious Field – Doing charity work, helping others.

Or, one could have an Aggressive Field – everyone fights with everyone else and the fittest survives.

All these Fields depends on the choice of the participants and not necessarily the designer. The design merely allows both Fields to exist.

Some of the rules we have seen in the Inanimate Field are also necessary for the Animate Field to exist as follows:

The huge mass that must collapse to create the spinning Black Hole which becomes the center of a new galaxy.

The mass of stars forming within the revolving gas and dust clouds within the new galaxy.

The fusion of elements in the core of stars as they go through processes of birth, youth, middle age, old age and death.

The distribution of the elements formed in the core of stars via nova and supernova explosions.

Mass giving heavenly bodies different escape velocities. On the star objects weigh kilograms while on a planet's surface the same object must weigh a lot less. On the planet's moon the same object would weigh even less.

Design

Heat travelling from hot surfaces to cool surfaces always trying to reach an equilibrium. This is the rule that causes volcanoes and hot springs on the Earth. It is necessary to form the igneous rocks and metamorphic rocks as seen earlier.

Pressure travelling from high to low. This rule is responsible for many of the explosive events. If two materials are sliding past each other and one is obstructed. The pressure can build up and the delayed release of energy results in an explosive event.

Elements are formed with a central nucleus of protons and neutrons with electrons revolving around. These rules create charges to build up allowing the atoms to form all manner of complex molecules and compounds.

The chemistry rules that allow two atoms of hydrogen to combine with one atom of oxygen to form the water molecule, a substance so essential for life on Earth.

Energy being transformed from one form to another. Nuclear energy in the star is converted into heat energy. Heat energy keeps water in liquid form on the Earth. Constant heat applied to liquid water gives it the energy to evaporate and become a gas. As it rises up energy is lost and it must fall back as rain. Water can be stored in a dam. Water released from an outlet in a dam can be used to drive a turbine. Stored energy in water is converted to mechanical rotational energy. The rotational energy can be used to generate electricity or electrical energy.

Energy changing the state of matter. Heat energy added to solid ice transforms the ice to liquid water making the particles more energetic. Heat energy added to water transforms the water into gaseous water vapour making the particles still more energetic. Conversely, when energy is removed from water vapour gas it becomes a liquid. When energy is removed from liquid water, it becomes a solid ice. This transfer of energy rule is true for nearly every substance. However, different substances have different melting and boiling points. According to this rule a planet or moons temperatures and pressures would determine the substances that can exist on its surface as solid, liquid and gas.

Generally, as a rule when large dust clouds within nebular collapse they form a star with planets revolving around them. A significant rule is that when objects collapse or shrink inwards the collapsed body of matter begins to spin. This simple rule is significant because it is able to start a new galaxy when a Black Hole is formed. The rotation rule is important because it gives us a sensation of time. It also causes the star, planets and moons to rotate when they are formed. It keeps everything active and in motion. This rule is very significant for life on Earth and is hence critical to the design of the system. If the Earth did not rotate, one side of the planet would always face the Sun and it would get extremely hot. Also, one side of the planet would be always facing away from the Sun and would get extremely cold.

The rules mentioned above exist within the Fields of Heaven and their existence impose order within the Fields creating the beautiful galaxies, stars, planets and moons.

The rules generally allow the galaxies, stars, planets and moons to be born, to live for relatively long periods of time and eventually to die. The rules allow the stars to create the elements and to provide a Field for its associated planets and moons. The rules allow the planets and moons to have varying temperatures, pressures and conditions which can give rise to all manner of inanimate matter and animate matter. The fact that the rules allowed human beings to arise on Earth and discover the Fields of Knowledge is remarkable. For the Religious Field it is an indication of the existence of a Creator. Rules must have a Rule Maker a planner to:

1) Make the dust form the shapes and structures of inanimate matter.
2) Make inanimate matter of atoms join together to form molecules and compounds.
3) Make animate matter in the form of cells and lifeforms.
4) Make cells with genes and chromosomes with codes to transmit and reproduce data relevant for the cell and lifeform.
5) Add complications to allow cells and lifeforms to have sensors and brains to find sunlight, food and water.

Design

6) Add complications to allow cells and lifeforms to digest food and remove waste products.
7) Create human beings with minds capable of discovering the existence of the Fields of Knowledge and the complications of the process of creating Mind from Matter.

Each of the Fields of Knowledge have a complex set of rules or criterion, which must be followed by the people within the Field of Knowledge. A human mind is a necessary component to be a part of a particular Field and usually many past human minds have been responsible for creating the Field. This shows the importance of Mind in creating rules for the Fields of Knowledge on the Earth and can be used to support the view for a Creator whose will is carried out in Heaven.

A simple example of rules and Fields on the Earth is the Transport Field. As the amount of cars increased it became increasingly necessary to have rules for safe road transport. Today, we have cars being driven along roads with traffic lights, speed signs, road signs and parking signs. The rules create order and the traffic flows until someone disobeys the rules causing an accident. The accident is a form of disorder or chaos.

It is interesting to note that even to write this book, I had to follow rules. I had to use the rules of the English language which has a 26 letter alphabet. The alphabet is used to construct words which have meanings when put into sentences. The sentences are used to form the chapters. The twelve chapters form the book. The book has to be bound together with a front and back cover before it is published. Here is the interesting part, when the book is published, I have no control of its meaning to the other minds that read the book. I am merely the author and with the publisher, we become the designer of the book. The meaning of the book will be different for different people sometimes dependent on the Fields in which they find themselves.

The Fields of Knowledge are significant in revealing the true nature of the design, both on the Earth and in Heaven as follows.

If a human being had no knowledge of the Fields of Knowledge and the person had the full use of his senses of smell, touch, taste, sight and hearing while in a modern city on the Earth, the person would experience a magical world where:

a) Houses had liquid water flowing out of taps.
b) Houses had electricity to operate many electrical devices.
c) Houses had natural gas that could be used for heating and cooking.
d) Houses had television, telephones and other communication systems.
e) Food was always available in huge supermarkets.
f) Cars, buses and trams were moving along streets taking people to their destinations.
g) Planes were flying in the air.
h) Ships and boats were sailing along the water of lakes and rivers.
i) Trains were moving along railway tracks.
j) Tall building were linked by extensive road and bridge networks.

The person seeing these events would feel that these services are all automatically provided because they exist and they make life more comfortable. The person might even think that some great designer made the system so that life was comfortable for human beings.

However, studying the Fields of Knowledge we know that the above systems did not just appear magically. All the above systems had to be designed, manufactured and implemented by many human minds using rules and standards developed in the many Fields of Knowledge as follows:

a) The liquid water flowing out of taps is part of the Water Engineering Field that includes plumbing.
b) The electrical power in houses is part of the Building Services Electrical Field.
c) The natural gas is part of the Building Services Gas Field.

DESIGN

- d) The televisions, radio and telephones are part of the Building Services Communication Field.
- e) The food available in supermarkets is part of the Marketing and Business Services Field.
- f) The cars, buses, planes, trains, ships and trams are manufactured in business companies and are part of a Manufacturing Field.
- g) The movement of the cars, buses, planes, trams, ships and trams are part of the Transportation Field.
- h) The tall building, roads and bridges are part of an Architectural Field and Civil Engineering Field.

Obviously, a combination of millions of living and dead human minds working in many Fields of Knowledge over long periods of time have produced what the person without any knowledge of the Fields of Knowledge sees as automatically provided by a mysterious designer.

Similarly, when looking at Heaven and Earth without the Fields of Knowledge one sees a magical world:

- a) A morning blue sky with the yellow sun shining as it moves across the sky.
- b) A night sky with stars and a white moon.
- c) Trees and plants growing out of the ground.
- d) Trees and plants with beautiful flowers and fruit.
- e) A variety of fish in the water.
- f) Water falling from heaven as rain.
- g) A variety of animals living on the land.
- h) A variety of insects and birds moving around in the air.
- i) A breathable atmosphere enjoyed by plants and animals.
- j) Water flowing in rivers and streams to lakes and oceans.

The design appears magical. The trees, plants, animals and birds need water and it is provided from heaven as rain throughout the year. The Sun appears for half of the day and heats up the place. Then, it disappears for the remainder of the day to allow the stars and moon to operate and provide light and no heat, as if to cool the place down and create an atmosphere for resting. The spring, summer, autumn and winter cycle allow for more variety which in many cases make the day and night cycle seem more magical and beautiful. Lifeforms respond to these changes in a variety of ways. Some trees shed their leaves in winter, perhaps to conserve energy, which is scarce as the Sun shines for a shorter time. Some animals hibernate in winter to conserve energy. The day and night cycle and the spring, summer, autumn and winter cycle are regular and continuous from year to year and from generation to generation. It is as if the design allows the creatures to get used to the conditions. When storms occur they usually last for a relatively short time before the normal cycles are restored. In a lifeforms relatively short lifetime there are relatively few major disasters and usually many good times for most lifeforms to experience a magical world.

However, when one looks at the magical world through the Fields of Knowledge, one sees the enormous time scale for the design to develop according to rules and laws:

a) The Black Hole Field had to form to start time for the galaxy.
b) The Galaxy Field had to form for the stars to come into existence.
c) The Star Field had to form to create the elements and to support the Planetary and Moon Fields.
d) The Planetary and Moon Fields had to form and cool, for inanimate matter and animate matter to be formed.
e) The Inanimate Fields had to form to create a surface where matter could exist as a solid, liquid and gas.
f) The Animate Fields had to form to allow cells to come into existence with sensors and a brain.
g) In the case of Earth human consciousness had to develop the Fields of Knowledge.

# Design

A tremendous design spanning billions of years. A design with billions of rules and laws. A Universe with billions of galaxies each with billions of stars and trillions of planets and moons. As we know the energy output of our ordinary star the Sun, we can understand the enormous energy output for the complete design. The cost of such a design is unimaginable and to think that it is given to us freely by the Universe and the Creator.

It is not surprising to find that we have a Religious Field that is full of praise and thanksgiving to the Creator for the Universe with its Fields containing the rules and laws to bring inanimate and animate matter into existence. Obviously, if the rules did not allow human consciousness to come into existence, no lifeform would know about the Fields that exist all around us and there would be no purpose to the Creator's design.

The significance of the Fields of Galaxies and the Star, and the Fields of Knowledge and the Human Mind, might indicate that Earth with human beings are the Center of Consciousness in the Universe. Through the Fields of Knowledge, human beings have been able to build machines and cities where millions of human beings can live in relative comfort. The cities are places where the Fields of Knowledge develop at an ever increasing pace creating many human stars in many Fields. This gives the design a purpose because it can be understood and utilized by many human minds.

# 12
# UTILISATION

The Fields all around us can best be described as a gift of the Universe and the Creator. To appreciate the Fields one must possess a human mind that has access to the Fields of Knowledge. The human mind can not only access the Fields of Knowledge but can utilize the Fields of Knowledge to create machines and cities that improve the human lifestyle.

One of the fundamental rules of the Creator is that energy can be transformed from one form to another. The Star Field generates enormous quantities of energy which are available in the Planetary and Moon Fields. We have seen earlier that solar energy was stored in the dead remains of plants and microscopic sea creatures as coal and oil deposits. Through the Fields of Knowledge we know that coal and oil are non-renewable sources of energy. In short, through the Fields of Knowledge human beings are able to generate large scale power supplies using non-renewable sources and renewable sources. The non-renewable sources include:

## Utilisation

1) Coal
2) Oil
3) Natural gas
4) Nuclear

The renewable sources include:

1) Solar
2) Wind
3) Hydro-electric
4) Tidal

Electricity generated at a power station has to be transported via sub-stations and cable networks to each individual consumer. Each consumer has fixed mains power.

Motor vehicles have engines that convert petrol or oil into energy to do work and move the vehicle.

Generators operating with diesel fuel are used as backup systems when fixed mains power fails or is unavailable.

Batteries produce energy via chemical reactions and they are self-contained and often used as a mobile energy source. Batteries are an extremely important means of powering machines when fixed power is unavailable.

The ability to utilize energy on a large scale is essential for operating the machines and the many cities on the Earth.

We have seen how the Inanimate Field of planets and moons follows the rules and transforms the molten material into a crust or surface with rocks and minerals as it cools over a relatively short period of time. Through the Fields of Knowledge human beings are able to find the rocks and minerals in the Earth's crust. Through the Jewelry Field and the Mining Field, the rocks and minerals are mined and processed to produce a wealth of elements, compounds and chemicals that are necessary to build machines and cities. Generally, cities require many machines, hence the large scale use of these resources are necessary to satisfy the needs of millions of lifeforms. It is essential to use the minerals and resources carefully. Use of renewable resources are essential for operating machines and cities over a long period of time.

We have seen how the Animate Field on the Earth produces biomes and ecosystems where plants and animals live in habitats and communities. Plants make their own food by absorbing nutrients from the soil and together with sunlight they produce energy which is distributed to the cells in their bodies via a liquid called sap. Animals eat plants and sometimes other animals to get the energy inputs which are distributed to the cells in their bodies via a liquid known as blood. The energy is needed by the lifeform to carry out body functions and to move around or to remain alive. Generally, plants provide food and shelter for animals. Animals help pollinate plants. Of all the lifeforms found on the Earth, human beings are the only beings to acknowledge the existence of a Creator. This led to a caring for the Creator's creations which led to a more settled existence with the raising of animals, birds and fish for food and the cultivation of the land to grow crops. These activities led to the development of the Fields of Knowledge. Generally, cities require the large scale use of food and water resources to feed millions of hungry lifeforms. It is essential to use the plant, fish, bird and animal resources carefully with love and care.

The modern world has seven continents Africa, Asia, Europe, Oceania, Antarctica, North America, and South America. Each continent has states controlled by governments. Each state has a number of cities that hold the resources to support usually millions of human beings and numerous animals and plants. All the people living in the cities are supported and have lifestyles generated by the numerous machines all around.

Machines are built to specifications using materials mined and processed from the Earth's rocks and minerals. Machines use the energy sources above to do work. However, machines use the energy resources described above that have been developed through the Fields of Knowledge. They are good because these resources do not depend on the exploitation of other lifeforms.

Utilisation

Machines assist human beings in carrying out work which includes:

1) Making bricks, cement and concrete to build houses, shops, educational institutions and offices where people can work in many of the Fields of Knowledge.
2) Making jewelry and clothing for people in the Clothing and Grooming Field.
3) Making materials for roads and the roads themselves for the Transportation Field.
4) Making shipping ports and train stations for the Transportation Field.
5) Making airports and runways for the Transportation Field to operate Efficiently.
6) Making gardens, parks and sporting facilities for human recreation and entertainment.

Buildings are provided with machines that generate water services, sewerage services, electricity services and communication services. Machines are everywhere within buildings we have lights, power outlets, television, computers, cooking appliances, air-conditioners, hair dryers, washing machines, dish washers, tooth brushes, clocks, and shavers. On the streets, we have motor vehicles of all varieties and shapes. On the waters and oceans, we have ships, boats and all manner of sailing ships. In the air, we have all types of planes. All these machines are carefully designed, constructed and installed by Human Beings working in the millions of Fields of Knowledge. Some of the Fields of Knowledge have been listed in chapter 9 of this book. All these Fields of Knowledge are used to enable a city to provide accommodation, medical, education, security, transportation, food and communications for millions of human beings, so that they can become stars and enhance the Fields of Knowledge.

Every design has a purpose, perhaps, one of the purposes of the design involving a Universe with Black Hole Fields, Galaxy Fields, Star Fields, Planetary and Moon Fields, Inanimate Fields and Animate Fields is to provide a means whereby animate lifeforms can understand the rules and laws of the Fields. To understand Mass and Escape Velocity, so that we can escape into Space with the Fields of Knowledge. With a knowledge of the Inanimate Fields we can build machines that can exist on many of the planets and moons in many galaxies. We can take our consciousness and the Fields of Knowledge to other stars and galaxies, so that all the galaxies will have matter and mind.

One of my previous books mentions the Yin Yang philosophy. The above represents the Yin side of the Fields of Knowledge. The Creator's design is an open system and obviously if there is an open system there must be a Yang side. The Yang side made clearly visible through the Fields of Knowledge is that, manufacturing machines on a large scale to satisfy the requirements of cities causes the following issues:

1) The biomes where lifeforms exist in a delicate balance are affected by loss of habitat when trees are cut down to make way for roads, dams, tunnels and the human city biome.
2) The use of fossil fuels releases large quantities of carbon into the atmosphere. Many scientists believe that this might cause catastrophic climate change.
3) Plants and animals are stressed by city lights, water infrastructure and roadways that cross their paths to food and water supplies.
4) Invasive species and bacteria can spread far more easily because human beings are capable of travelling all around the world very quickly through the efficient Transportation Field.

The Fields of Knowledge offer solutions, however the solutions are usually very costly and difficult to carry out. Some of these solutions are:

1) The reduction in the use of fossil fuels and the use of solar, water and wind power on a world wide scale.
2) Creation of rules and use of technology to reduce the spread of invasive species and harmful bacteria.

## Utilisation

Perhaps as a general rule, we can stop our needless preying on plants and animals in their biomes. Instead we can pray to the Creator to help us improve the Fields of Knowledge with brighter stars, so that we can manufacture better machines that will help us to build better and greener city biomes on the Earth that will exist in harmony with the biomes on the Earth. This will certainly make us responsible and enable us to take our consciousness and the Fields of Knowledge to other stars and other galaxies.

We all know that everything in the Universe made by the Creator goes through processes of birth, youth, middle age, old age and death. This includes our galaxy and our star, the Sun. Maybe the above is part of an overall design plan. By being responsible and able to escape from Earth's mass we will ensure that our consciousness will survive elsewhere in the Creator's Universe, so someone on another star or in another galaxy will know about the beautiful Fields of Knowledge in the Creator's Universe after our star ceases to exist.

# ABOUT THE AUTHOR

Vincent J. Hyde was born in September 1954 in Calcutta, India. While living in Calcutta, He studied in Saint Xavier's College.

He immigrated to Sydney, Australia, in February 1970. While living in Sydney, he completed his secondary education at Marcellin College and Merrylands High School, where he obtained his Higher School Certificate in 1973.

He completed the electrical engineering degree course at the University of New South Wales in 1979. He earned a postgraduate diploma in illumination design at the University of Sydney in 1983.

He worked as an electrical engineer at the New South Wales Public Works from 1979 to 2014 and retired from active engineering duties in 2014. Since 2014, he has been writing books to continue his professional development.

He has written five books with the publisher Balboa Press, as follows:

1. *Heaven and Earth*
2. *A Journey from Dust to Consciousness*
3. *A Message from the Neighbours*
4. *Earth's Reply*
5. *The Alien World*

He is a current member of the Institution of Engineers Australia (MIEAust), Chartered Professional Engineer (Ret) No. 1387147.

www.ingramcontent.com/pod-product-compliance
Lightning Source LLC
Chambersburg PA
CBHW020441220526
45464CB00002B/799